Roya Ahmadi

Group Theory in Inorganic Chemistry

Roya Ahmadi

Group Theory in Inorganic Chemistry

Group Theory

Noor Publishing

Imprint
Any brand names and product names mentioned in this book are subject to trademark, brand or patent protection and are trademarks or registered trademarks of their respective holders. The use of brand names, product names, common names, trade names, product descriptions etc. even without a particular marking in this work is in no way to be construed to mean that such names may be regarded as unrestricted in respect of trademark and brand protection legislation and could thus be used by anyone.

Cover image: www.ingimage.com

Publisher:
Noor Publishing
is a trademark of
International Book Market Service Ltd., member of OmniScriptum Publishing Group
17 Meldrum Street, Beau Bassin 71504, Mauritius
Printed at: see last page
ISBN: 978-620-0-77534-4

Group Theory in Inorganic Chemistry

Dr. Roya Ahmadi

Department of Chemistry, Yadegar-e-Imam Khomeini (RAH) Shahre-rey Branch, Islamic Azad University, Tehran, Iran

Email: roya_ahmadi_chem@yahoo.com

Dr. Roya Ahmadi

Department of Chemistry, Yadegar-e-Imam
Khomeini (RAH) Shahre-rey Branch, Islamic Azad
University, Tehran, Iran

Email: roya_ahmadi_chem@yahoo.com

Dr. Roya Ahmadi began her undergraduate studies in general chemistry in 1992 at Shahid Beheshti University of Tehran. Then, from 1997 to 2000, she completed her master's degree in inorganic chemistry at Kharazmi University of Tehran (teacher training). In 2006, she received her PhD in Mineral Chemistry from the Islamic Azad University of Science Research Branch. In the same year, she worked as an assistant professor at Islamic Azad University, Yadegar Branch of Imam Khomeini, and was promoted to associate degree in 2014. The result of a part of her research activities in the fields of empirical and computational chemistry over the years, more than 40 ISI articles, 14 scientific-research articles, and 14 specialized scientific articles, 161 posters and lectures at national and international conferences and guidance and counseling. of more than 70 postgraduate and doctoral students in theses. Also, as a member of the National Standards Committees, she has collaborated with the National Iranian Standards Organization to develop and prepare 32 national standards in the field of chemical and medical engineering. This simple and comprehensible text is the result of years of teaching group theory in advanced and inorganic chemistry at undergraduate and graduate chemistry levels.

This Book is dedicated to

To my kind Parents,

My Devoted Wife

&

My Dear Girls *Sarah* and *Sahar*

Table of Content

Chapter III

Exercise

Chapter IV

Application of group theory

Chapter V

Conclusion

References

Introduction

Group (group) is one of the most important algebraic structures that plays a fundamental role in abstract algebra and is very important in various sciences such as crystal, physics, and quantum and so on. The idea of forming a theory of vortices was formed when mathematicians observed that the structures that study can be found in common properties, and if they can study all of these properties for a certain structure, they actually study a large part of the same structures, thus saving time. The branch of mathematics that is devoted to reading is called group theory. The theory of vortices was developed by major composers of classical algebra, numbers theory, geometry and analysis. The classic algebra was made in 1984 by the works of joseph louis lagrange on the multi - sentence equations. However , euler , lagrange , lagrange , lagrange , lagrange , lagrangian , and french mathematician were the first to investigate the field theory. Especially because of his basic theorem, which is an interface between the ring and the ring, and today it is considered as the reader. The first application of deconvolution in describing the effect of root hairs is a multi - term equation which is used by Lagrange, which is based on this approach. He discovered that the roots of all the cases he tried were functions of their corresponding roots. After him, attempts to prove the absence of a direct solution to the solution of fifth and higher order equations were taken into account. After him, he published his first work on his theory at the age of eighty - five. However, his contribution was not considered before the publication of the collection of articles in this year. Until then there were no clear principles for the definition of the group. In 2008, kelly gave the first principles for the issue, but his definition soon fell short of value. In 2001, The Hague renewed the principle of topical application. Weber was the first to define the infinite domain and define infinite domain for infinite domain. Walter van dyke first presented the first modern definition of the group. The aim of this study was to investigate the relationship between self - efficacy and self - efficacy in both groups. Today's theory has become the most fundamental theory in abstract algebra and is a source of research for mathematicians.

Chapter I

Symmetry

Symmetry

The symmetry of a geometric shape means that one part of it is the same as the other part of it, such as the verse image; in other words, if rotation at a certain angle does not alter the appearance of the shape, the shape is symmetric.

Figure 1. Symmetry in nature

Elements of Symmetry

1. Symmetry plate (σ)

2. Center of Symmetry (i)

3. General axis (C_n)

4. Axis of compound period (S_n)

5. Same element (E) or (C_1)

Each symmetric element is ascribed to a symmetric action, which we will explain in the following:

1. Symmetry plate (σ): A plate that divides a shape or molecule into two equal parts that mirror the image of each other and reflect the type of symmetrical action.

Each planar molecule has a plane of symmetry which is the plane of the molecule.

Molecule Plate: a plate that divides the molecule into two diameters and is always present in planar molecules. In the form of multiple axes with different degrees, we consider the axis that has the largest order as the principal axis, and we classify the symmetrical plates according to the principal axis as follows:

A)σ_h: Horizontal reflection plate which is perpendicular to the main rotation axis (C_n).

B) σ_v: Vertical reflection plate that covers the main rotation axis (C_n).

C)σ_d: If there are multiple axes (C_2) in the figure and the plane of symmetry is the angle between two axes (C_2), they are called that plane(σ_d).

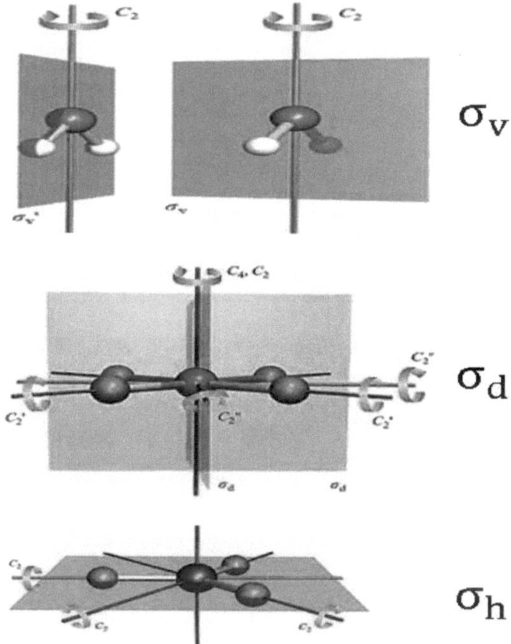

Figure 2. Show of Symmetry plate in H_2O (σ_v). $PtCl_4$ (σ_d), BF_3 (σ_h)

2. Center of symmetry (i): The act of inversion is the center of symmetry, meaning that if a point exists in the shape and is extended, it must reach the same point on the other side of the figure, so we say the shape has a center of symmetry.

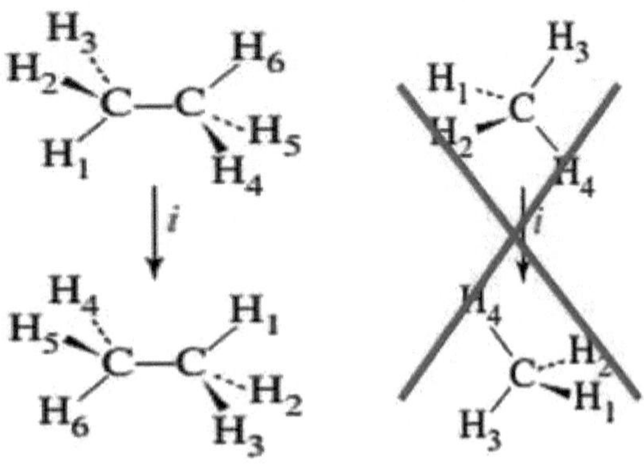

Figure 3. Check of Center of symmetry (i) in C_2H_6 and CH_4

3. Proper rotation axis (C_n): The proper rotation axis is a line that if we rotate the molecule as much (2π / n) around it, the molecule reaches a similar state. The proper rotation axis(C_n) mark is which n is called degree axis. In the figure where there are several axes with different degrees, we consider the axis with a higher degree to be the proper rotation axis (Note: As mentioned earlier, the symmetrical pages are classified according to the principal axis).

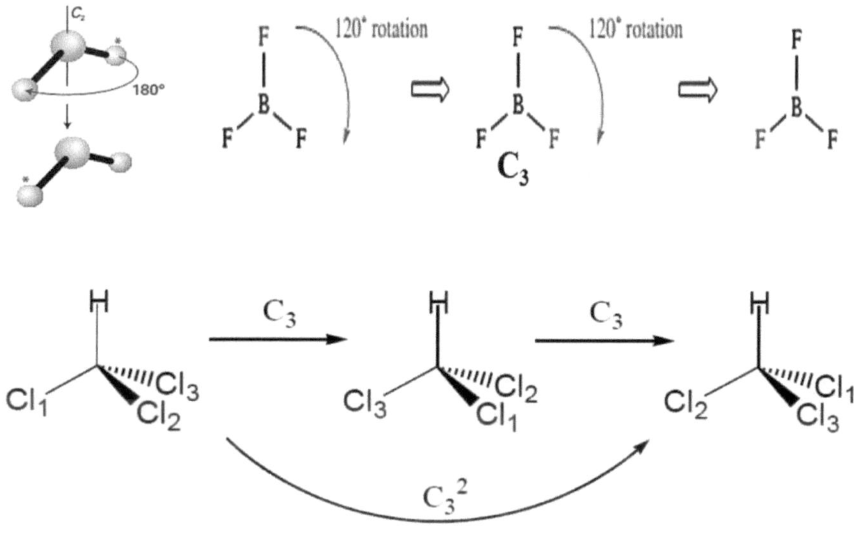

Figure 4. Proper rotation axis (C_2) in H_2O and (C_3) in BF_3 and $CHCl_3$

Note: In regular shapes, the main rotation axis will be as large as the number of sides; for example, triangle C_3, hexagon C_5, hexagon C_6, etc. In group theory, order and power can be simplified, for example:

$$C_3^3 = C_1 = \mathbf{E}$$

$$C_4^2 = C_2$$

In the proper rotation axis, the more the degree, the angle becomes smaller:

$$C_2 \qquad \frac{2\pi}{n} = \frac{2\pi}{2} = 180°$$

$$C_3 \qquad \frac{2\pi}{n} = \frac{2\pi}{3} = 120°$$

$$C_4 \qquad \frac{2\pi}{n} = \frac{2\pi}{4} = 90°$$

$$C_5 \qquad \frac{2\pi}{n} = \frac{2\pi}{5} = 72°$$

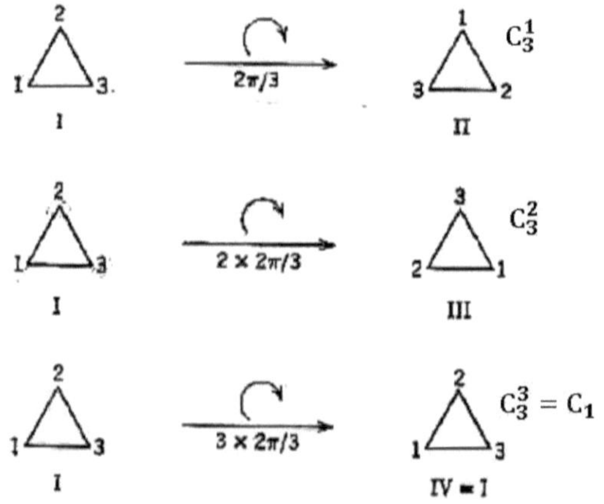

Figure 5. Proper rotation axis $(C_3^1, C_3^2, C_3^3 = C_1)$ in Triangle

Compound rotation axis (S_n): This axis consists of two operations: a) General rotation (C_n) b) Symmetrical plate (σ) which is perpendicular to the (C_n) axis. If the molecule C_n and the plane of reflection perpendicular to it exist separately, then there will be S_n; but in some cases neither of these two elements may exist alone, but the compound rotation axis (S_n) can exist. For example, a quadratic arranged cubic arrangement will be considered whose vertices are filled in: all regular tetrahedral molecules that have SP^3 hybridization such as methane (CH_4), Silane (SiH_4), nickel tetra carbonyl (Ni(CO)$_4$) are part of the T$_d$ point group with three C$_2$ axes, four C$_3$ axes and six σ_d plates. Now we can first apply the σ plane and then apply C$_4$: So as you can see there is a combination of C$_4$ and σ_h called S$_4$:

$$C_4\ \sigma_h = S_4 = \sigma_h\ C_4$$

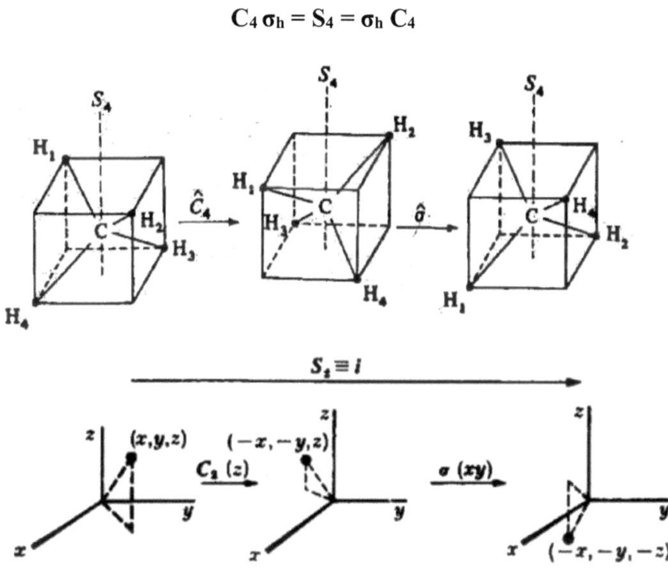

Figure 6. Compound rotation axis (S_n), S$_4$ in CH$_4$ and S$_2$=i

Identical Element (E): an element defined by symmetry (C_1). This means that if we rotate the molecule as much as 360° around any arbitrary axis, it will return to the same initial state.

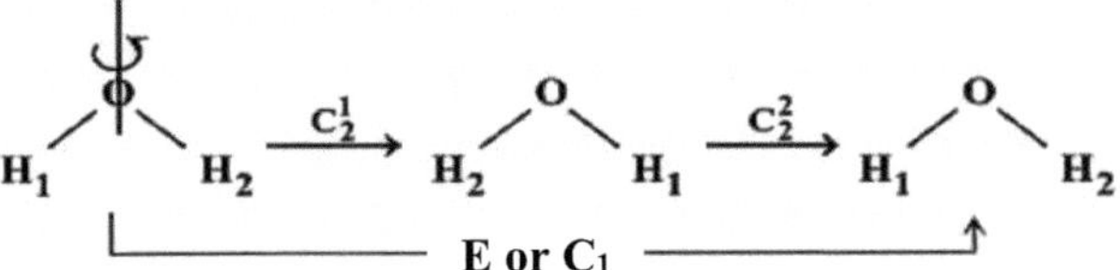

Figure 7. Identical Element (E) is C_1

Each symmetric operation, once performed, brings the molecule or shape back to the same state as the initial state, and after several repetitions the state will be exactly the same as the original state.

Consider the number of steps for different symmetric elements:

Symmetry plate: Consider the water molecule

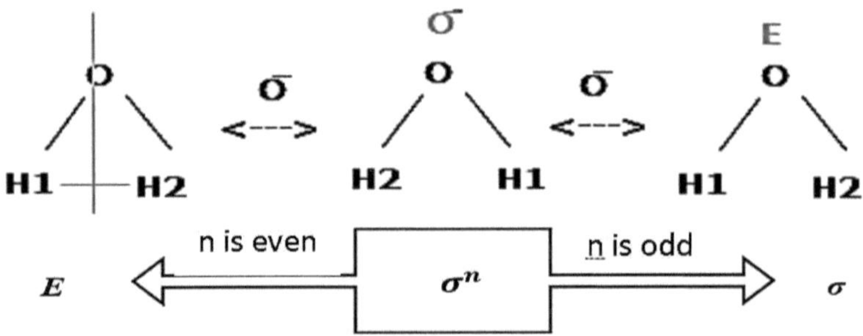

Figure 8. Symmetry plate after several repetitions

Center of Symmetry

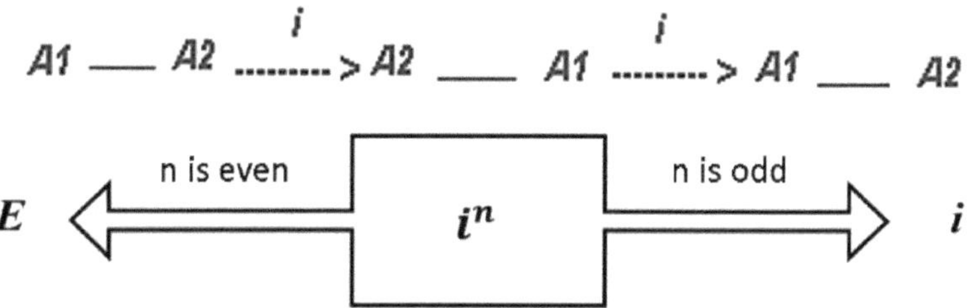

Figure 9. Center of Symmetry after several repetitions

General rotation axis: If we repeat C_n, n times we get the same element:

$$C_n^n \xrightarrow{\textit{If n be even or odd}} C_n^n = E$$

Compound rotation axis

Investigating S_4:

$S_4^1 = C_4 \cdot \sigma_h = S_4^1$

$S_4^2 = C_4^2 \cdot \sigma_h^2 = C_2 \cdot E = C_2$

$S_4^3 = C_4^3 \cdot \sigma_h^3 = C_4^3 \cdot \sigma_h = S_4^3$

$S_4^4 = C_4^4 \cdot \sigma_h^4 = E \cdot E = E$

$S_4^5 = C_4^5 \cdot \sigma_h^5 = C_4^4 \cdot C_4^1 \cdot \sigma_h = E \cdot S_4^1 = S_4^1$

Investigating S_5:

$S_5^1 = C_5^1 \cdot \sigma_h^1 = S_5^1$

$$S_5^2 = C_5^2 \cdot \sigma_h^2 = C_5^2 \cdot E = C_5^2$$

$$S_5^3 = C_5^3 \cdot \sigma_h^3 = C_5^3 \cdot \sigma_h = S_5^3$$

$$S_5^4 = C_5^4 \cdot \sigma_h^4 = C_5^4 \cdot E = C_5^4$$

$$S_5^5 = C_5^5 \cdot \sigma_h^5 = E \cdot \sigma_h = \sigma_h$$

$$S_5^6 = C_5^6 \cdot \sigma_h^6 = C_5^6 \cdot C_5^1 \cdot E = E \cdot C_5^1 \cdot E = C_5^1$$

$$S_5^7 = C_5^7 \cdot \sigma_h^7 = C_5^5 \cdot C_5^2 \cdot \sigma_h = E \cdot C_5^2 \cdot \sigma_h = S_5^2$$

$$S_5^8 = C_5^8 \cdot \sigma_h^8 = C_5^5 \cdot C_5^3 \cdot E = E \cdot C_5^3 \cdot E = C_5^3$$

$$S_5^9 = C_5^9 \cdot \sigma_h^9 = C_5^5 \cdot C_5^4 \cdot \sigma = E \cdot S_5^4 = S_5^4$$

$$S_5^5 = C_5^5 \cdot \sigma_h^5 = E \cdot \sigma_h = \sigma_h$$

$$S_5^{10} = C_5^{10} \cdot \sigma_h^{10} = C_5^5 \cdot C_5^5 \cdot E = E \cdot E \cdot E = E$$

$$S_5^{11} = C_5^{11} \cdot \sigma_h^{11} = C_5^5 \cdot C_5^5 \cdot C_5^1 \cdot \sigma_h = E \cdot E \cdot S_5^1 = S_5^1$$

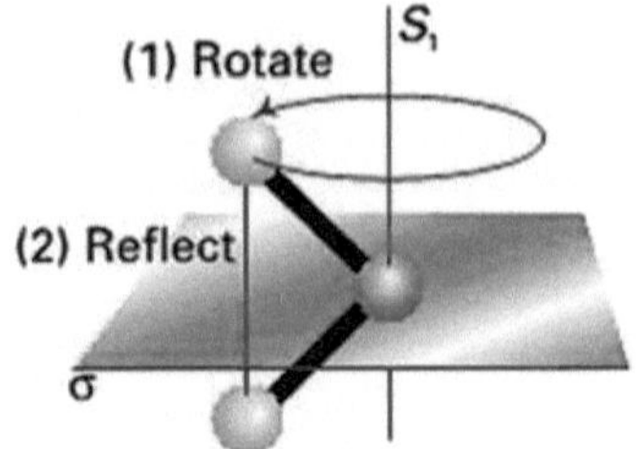

Figure 10. Compound rotation axis after several repetitions

Exercise 1: Prove $\sigma = S_1$.

$$S_1 = C_1 \cdot \sigma_h = E \cdot \sigma_h = \sigma_h$$

Exercise 2: Prove that the function S_n^n if n is even, is the value of E, and if n is odd also is the value of σ_h?

$$S_n^n \xrightarrow{\ n\ even\ } = C_n^n . \sigma_h^n = E . E = E$$

$$S_n^n \xrightarrow{\ n\ odd\ } = C_n^n . \sigma_h^n = E . \sigma_h = \sigma_h$$

$$S_1 = \sigma$$
$$S_n^n = \sigma_h \quad \text{if } n \text{ is odd}$$
$$S_n^n = E \quad \text{if } n \text{ is even.}$$

Exercise 3: Prove that the action S_{2n}^n if n is even, is the equivalent of C_2, and if n is odd is also the value of i?

$$S_{2n}^n \xrightarrow{\text{n even}} = C_{2n}^n . \sigma_h^n = C_2 . E = C_2$$

$$S_{2n}^n \xrightarrow{\text{n odd}} = C_{2n}^n . \sigma_h^n = C_2 . \sigma_h = S_2 = i$$

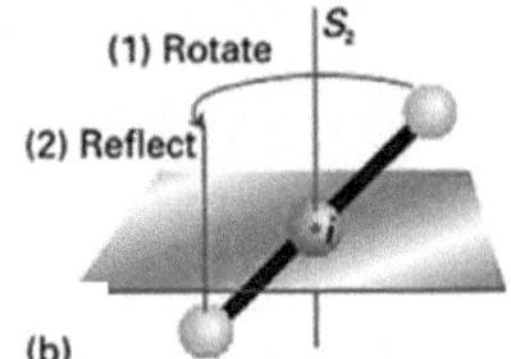

Symmetry group or point group

Symmetry group: The set of symmetry operations performed on a molecule is called the point group or the symmetry group of that molecule. The set of these actions does not move the molecule from one point to another but leaves the molecule in the same space.

Group theory: It is mathematical theory, for a set of elements between which there is a definite binary relation; the relation between symmetry operations found in a point group or symmetric group can be obtained by means of group theory; Familiarity with group theory is required.

Group: A set of elements if they form a group that meet the following four conditions:

$$\textbf{a.b = c} \qquad \textbf{b.a = c}$$

Law of Participation in Multiplication: The element that is the product of (a.b) If multiplied by element c, the product of the element is the same as can be obtained by multiplying element a by the product of (b.c).

(a.b). c = a. (b.c)

Identical element: Each group must have at least one element represented by the symbol E, called the same element.

E. a = a. E = a

The inversion element (a^{-1}): The property of the inverse element is that if we multiply the element by its inverse, we reach E; that is, for every element in a group there must be an element a-1 which holds the following relation:

a. a^{-1} = E

Subgroup: The smaller groups found in each group are called subgroups of that group. The subgroup has all the characteristics of a subgroup first;

h = k g *(k= constant)*

Sub typical Subgroup: Element E and the subgroup itself are subgroups of subgroups (meaning they must exist).

Conjunctive Elements: When $XAX^{-1} = B$ is a relation between two elements A and B, these two elements fall into a group and B is called the similarity transform of A.

Classes of Group: A set of conjugate elements of a group is called a class or group class.

$$E^{-1}EE = EEE = E$$
$$A^{-1}EA = A^{-1}AE = E$$
$$B^{-1}EB = B^{-1}BE = E$$

$$E^{-1}DE = D$$
$$A^{-1}DA = F$$
$$B^{-1}DB = F$$
$$C^{-1}DC = F$$
$$D^{-1}DD = D$$
$$F^{-1}DF = D$$

$$E^{-1}AE = A$$
$$A^{-1}AA = A$$
$$B^{-1}AB = C$$
$$C^{-1}AC = B$$
$$D^{-1}AD = B$$
$$F^{-1}AF = C$$

Abelian group: A group where all multiplicative operations are interchangeable and each conjugate element is inverted; in other words, each class category contains one element. So in the multiplication table, all the elements on the original diameter are the same element and all the elements on the sides of this diameter are symmetrical.

Group rank (h): The number of members of a given group is called the rank of a group that is conventionally represented by the symbol h.

Multiplication Table: The multiplication table of a given group consists of h rows and h columns, and the order of the group members is not in a duplicate row or column.

Group 1: This group has only the same element.

C_1	E
E	E

Group 2: This group is a ring group.

C_2	E	A
E	E	A
A	A	E

C_s	E	σ_h
E	E	σ_h
σ_h	σ_h	E

C_i	E	I
E	E	i
i	i	E

Group 3: This group is also a ring group.

Cyclic Group: A group consisting only of the element X and all its exponentiations from X^h= E to X^2/group order).

For example, if we assume the following:

$X^1 = A$

$X^2 = A. A = B$

$X^3 = A. A. A = A. B = E$

$B. B = A. A. A. A = A. E = E$

Exercise: Write the multiplication table of the third order group.

C_3	E	A	B
E	E	A	B
A	A	B	E
B	B	E	A

$C_3 = C_3^1 \quad C_3^2 \quad C_3^3 \quad E$

$C_3^2 . C_3^2 = C_3^4 = C_3^3 . C_3^1 = E. C_3^1 = C_3^1$

C_3	E	C_3^1	C_3^2
E	E	C_3^1	C_3^2
C_3^1	C_3^1	C_3^2	E
C_3^2	C_3^2	E	C_3^1

Group 4

$C_4 = C_4^1 \quad C_4^2 = C_2 \quad C_4^3 \quad C_4^4 = E$

C_4	E	C_4^1	C_2	C_4^3
E	E	C_3^1	C_2	C_4^3

C_4^1	C_4^1	C_2	C_4^3	E
C_2	C_2	C_4^3	E	C_4^3
C_4^3	C_4^3	E	C_4^1	C_2

$$C_4^2 \cdot C_4^3 = C_4^5 = C_4^4 \cdot C_4^1 = E. \ C_4^1 = C_4^1$$

$$C_4^3 \cdot C_4^3 = C_4^6 = C_4^4 \cdot C_4^2 = E. \ C_4^2 = C_4^2 = C_2$$

The second type of group four is the group with each element upside down. The groups C_{2v}, C_{2h}, D_2 are of the following type:

Group C_{2v}: This group has all the properties of an abelian group, meaning that the product of each member of the group will have the same element in itself.

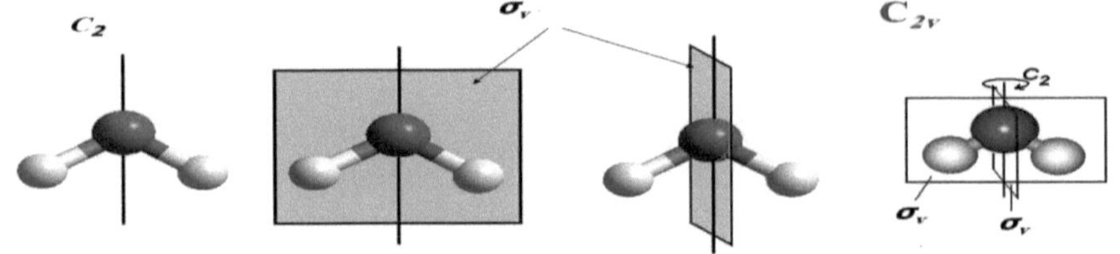

Figure 11. Elements of Symmetry in H_2O in Group C_{2v}

❖ **Note:** Whenever an element operates on the z axis in the Cartesian coordinate system (x, y, z), the x and y marks do not change but the z mark will not change.

$$C_{2v} = E \quad C_2 (z) \quad \sigma_{xz} \quad \sigma_{yz}$$

C_{2V}	E	$C_2(z)$	σ_{xz}	σ_{yz}
E	E	$C_2(z)$	σ_{xz}	σ_{yz}
$C_2(z)$	$C_2(z)$	E	σ_{yz}	σ_{xz}
σ_{xz}	σ_{xz}	σ_{yz}	E	$C_2(z)$
σ_{yz}	σ_{yz}	σ_{xz}	$C_2(z)$	E

As mentioned in the preceding pages, the elements are the same diameter of the original element, and all the elements on either side of this diameter are symmetrical.

High multiplication table proof:

$C_2\ (z) \cdot C_2\ (z) = C_2\ (z) \cdot C_2\ (z)\ (x,y,z) = C_2\ (z)\ (x^-,y^-,\ z) = (x,y,z) = E$

$C_2\ (z) \cdot \sigma_{xz} = C_2\ (z) \cdot \sigma_{xz}\ (x,y,z) = C_2\ (z)\ (x,y^-,z) = (x^-,\ y,z) = \sigma_{yz}$

$\sigma_{xz} \cdot \sigma_{xz} = \sigma_{xz} \cdot \sigma_{xz}\ (x,y,z) = \sigma_{xz}(x,y^-,z) = (x,y,z) = E$

$\sigma_{xz} \cdot \sigma_{yz} = \sigma_{xz} \cdot \sigma_{yz}\ (x,y,z) = \sigma_{xz}\ (x^-,y,z) = (x^-,y^-,z) = C_2\ (z)$

Practice: Write the multiplication table of the C_{2h} group.

(This group's reminder is 2n, so it will have 4 members.)

$C_{2h} = E \quad C_2\ (z) \quad i \quad \sigma_h$

Method of Obtaining C_{2h} Group Multiplication Elements

$C_2\ (z) \cdot i = C_2\ (z) \cdot i\ (x,y,z) = C_2\ (z)\ (x^-,y^-,\ z^-) = (x,y,z^-) = \sigma_h$

If we select the axis of action C_2 from the z axis of the coordinate machine:

$C_2(z) \cdot \sigma_h = C_2(z) \cdot \sigma_{xy}(x, y, z) = C_2(z)(x, y, z\text{-}) = (x\text{-}, y\text{-}, z\text{-}) = i$

If we select the C_2 action axis as the y axis of the coordinate machine:

$C_2(y) \cdot \sigma_h = C_2(y) \cdot \sigma_{xz}(x, y, z) = C_2(z)(x, y\text{-}, z) = (x\text{-}, y\text{-}, z\text{-}) = i$

So in any case, the result will be i.

C_{2h}	E	$C_2(z)$	i	σ_h
E	E	$C_2(z)$	σ_{xz}	σ_h
$C_2(z)$	$C_2(z)$	E	σ_h	i
i	i	σ_h	E	$C_2(z)$
σ_h	σ_h	i	$C_2(z)$	E

Note: For groups like C_{2h} D_2 D_{4h} O_h that have the center of inversion, use the g (even) and u (odd) subgroups.

Exercise: Write the multiplication table D_2.

D_n: C_n , $nC_2 \perp C_n$

$D_2 = E \quad C_2(x) \quad C_2(y) \quad C_2(z)$

$C_2(x(\cdot C_2(y)) = C_2(x(\cdot C_2(y)(x,y,z) = C_2(x)(x\text{-}, y, z\text{-}) = (x\text{-}, y\text{-}, z) = C_2(z)$

D_2	E	$C_2(x)$	$C_2(y)$	$C_2(z)$
E	E	$C_2(x)$	$C_2(y)$	$C_2(z)$
$C_2(x)$	$C_2(x)$	E	$C_2(z)$	$C_2(y)$
$C_2(y)$	$C_2(y)$	$C_2(z)$	E	$C_2(x)$
$C_2(z)$	$C_2(z)$	$C_2(y)$	$C_2(x)$	E

Exercise: Write the multiplication table of the fifth order group.

$$C_5 = C_5^1 \quad C_5^2 \quad C_5^3 \quad C_5^4 \quad C_5^5 = E$$

$$C_5^4 \cdot C_5^4 = C_5^8 = C_5^5 \cdot C_5^3 = E. \, C_5^3 = C_5^3$$

C_5	E	C_5^1	C_5^2	C_5^3	C_5^4
E	E	C_5^1	C_5^2	C_5^3	C_5^4
C_5^1	C_5^1	C_5^2	C_5^3	C_5^4	E
C_5^2	C_5^2	C_5^3	C_5^4	E	C_5^1
C_5^3	C_5^3	C_5^4	E	C_5^1	C_5^2
C_5^4	C_5^4	E	C_5^1	C_5^2	C_5^3

Point Groups

1. Group C_n : The point group C_n is obtained by repeating the symmetry function C_n; this group has the order n (in other words, the elements in the group = n).

If no symmetric element is found in the molecule, the point group will be the C_1 molecule, which is the most asymmetric form. The point of this group is to have a dipole moment and optical activity.

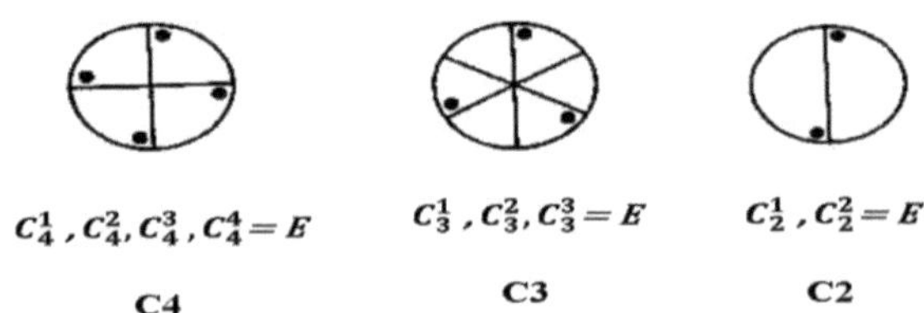

Figure 12. Highlighted images for point Groups C_n

Point group C_{nh}: If a point group is added to C_n in addition to C_n (σ_h), the group C_{nh} will be created). The order of this group is 2n (that is, if n = 3 then this group will have 6 members).

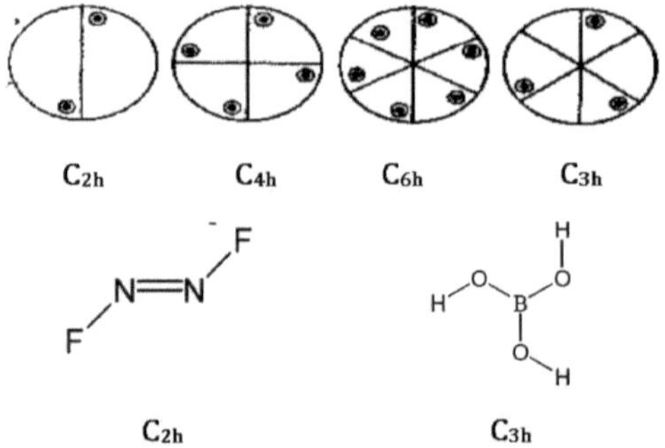

Figure 13. Highlighted images for point Groups C_{nh} and two molecules examples

- ❖ **Note:** There is no dot group called C_{1h} and it is represented as C_s, so if there is a symmetry plane in the molecule or shape, the dot group will be C_s.
- ❖ **Note:** In the C_{nh} point group, in addition to C_n and σ_h, there is also the symmetry element S_n, and if n is even, in the C_{nh} point group there will also be the symmetry element i: $C_2 \ \sigma_h = S_2 = i$.

3. Point group C_{nv}: In this point group in addition to having a rotating axis C_n , n there are n plane σ_v ; the order of this point group is 2n (for example if n = 2 then point group C_{2v} It will have 4 members.

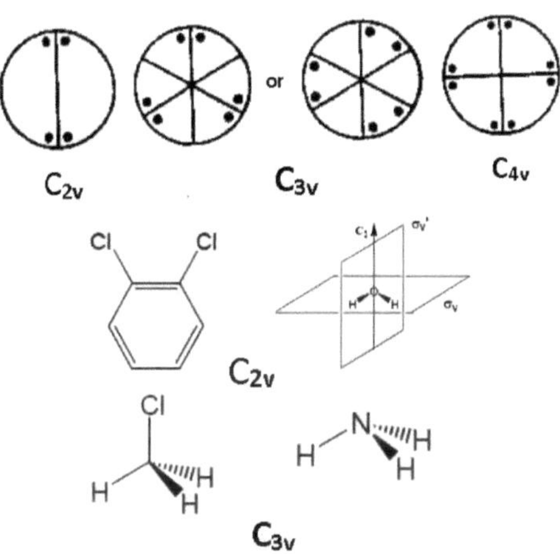

Figure 14. Highlighted images for   point Groups C_{2v} and four   molecules examples

4. Point group (D_n): If in a shape or molecule there is a rotating axis C_n , n besides the rotating axis C_2 perpendicular to the original C_n, the resulting point group will be D_n. The molecules belonging to this point group have optical activity and the order of this point group is 2n.

$$D_n\colon\ C_n\ ,\ nC_2^{\perp}\ C_n$$

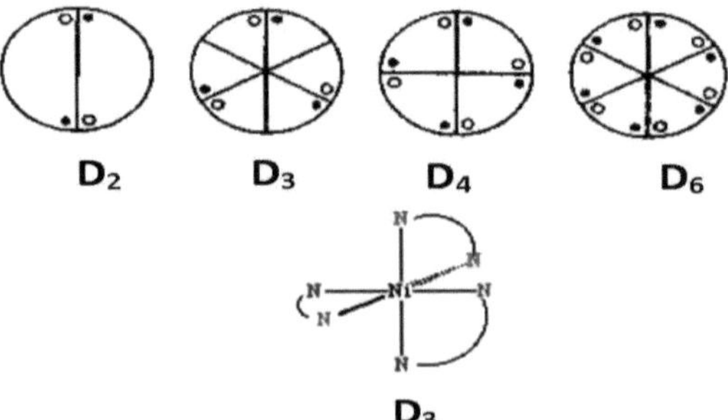

D₂　　**D₃**　　**D₄**　　**D₆**

D₃

Figure 15. Highlighted images for   point Groups Dₙ and one molecule example

5. Point groupD_{nh}: If there is a plane σ_h in addition to the symmetric elements

of the $D_n$group in a shape or molecule, the resulting point group will be D_{nh}. The order

of this group is a point 4n which also has the symmetry element i if n is even.

$$D_{nh}: C_n \, , \, nC_2 \perp C_n \, , \, \sigma_h$$

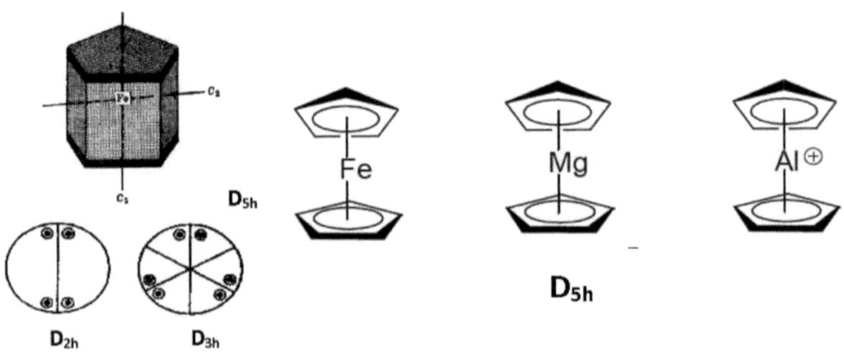

Figure 16. Highlighted images for point Groups Dₙₕ and molecule examples

In this point group, besides the σ_h page, there may be other pages for one shape. for

example:

$$D_{3h}: C_3 \, , \, 3C_2 \perp C_3 \, , \, \sigma_h \, , \, 3\,\sigma_v$$

6. Point group D_{nd}: If in a form or molecule there is also σ_d the D_n ، n plane in addition to the symmetric elements of the group D_n, the resulting point group will be D_{nd}. The order of this group is a point 4n which, if n, is also a symmetric element i.

$$D_{nd}: \; C_n \, , \, nC_2 \perp C_n \, , \, n\sigma_d$$

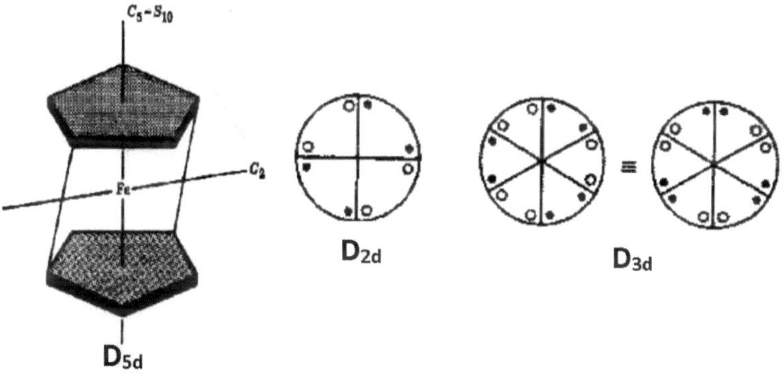

Figure 17. Highlighted images for   point Groups D_{nd} and   molecule example

❖ **Note:** If there are two halves in one molecule relative to each other, the molecule belongs to the D_{nd} point group.

Point groups $D_{\infty h}$ and $C_{\infty v}$: These two point groups are linear molecules. In both groups, the rotational axis of the molecule is C_∞, resulting in a zero degree of rotation.

If the linear molecule had a center of symmetry and the C_2 axes were perpendicular to the main axis, then the point group is $D_{\infty h}$, otherwise the molecule would have the point group $C_{\infty v}$.

Figure 18. Point Groups for linear Molecule examples

❖ **Note:** In these groups there is only one main axis, but in some groups there are multiple C_2 axes.

Cubic groups: There are molecules that have multiple axes above degree C_2, these are polyhomes whose axes are perpendicular to their faces; they are called cubic groups, such as T_d, O_h, O, T, T_h,

Group O and O_h: Group O is a general rotational group with only rotational axes and its rank is 24. By adding σ_h or i to the general rotational group, the point group O_h is obtained (octahedral), the order of which is 48:

O_h: $3C_4$, $3C_2$

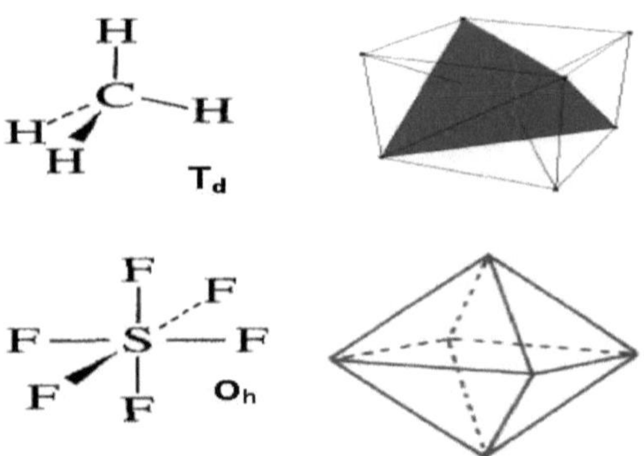

Figure 19. Point Groups T_d and O_h and Molecule examples

Dipole Moment

Molecules that have a permanent dipole moment are capable of rotating polarized light. The dipole moment is represented by the μ symbol.

- If $\mu_x = \mu_y = \mu_z = 0$, the molecule has no permanent dipole moment.
- If only one of the axial components of the non-zero dipole moment is non-zero, the molecule has a constant dipole moment and is oriented in the same direction.
- If the two axial components of the dipole moment are nonzero, the permanent dipole moment will be placed on a coordinate plane.
- ❖ **Note:** If the molecule has a center of symmetry (such as 1, 4-difluoro 2 and

5-dichloro benzene), the molecule will not have a permanent dipole moment.

Optical activity

This property is related to molecules that are incompatible with their mirror images, that is, in the case of compound-axis (S_n) optical activity:

Molecules with these dot groups are not optically active:

$$S_1 = C_1 . \sigma_h = \sigma$$

$$S_2 = C_2 . \sigma_h = S_2 = i$$

Sharp: Platonic bodies: those which are specific to the following all of their sides consist of regular polygons and equivalents All vertices and sides are the same.

The equilateral triangle is a quadrilateral that has 4 sides, 4 heads and 6 sides

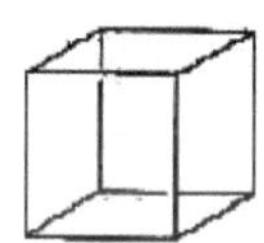

Cube square has 6 faces, 8 heads and 12 sides

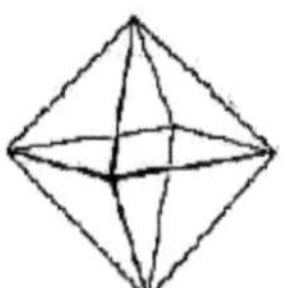

An octagonal equilateral triangle It has 8 faces, 6 heads and 12 sides

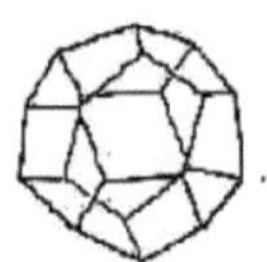

Pentagon ears has 12 faces, 20 heads and 30 sides

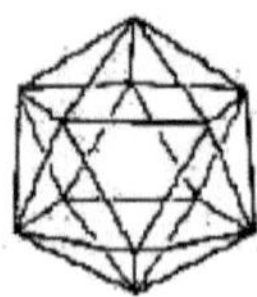

Equilateral triangle has 20 faces, 12 heads and 30 sides

Figure 20. Platonic bodies

Chapter II

Group Theory

Matrix

The representation of numbers or symbols in a $\begin{bmatrix} \\ \end{bmatrix}$ within a matrix is called a matrix. Each matrix consists of i rows and j columns, represented by the symbol Uij:

$$A_{m\times n} = [a_{ij}]_{m\times n} = \begin{bmatrix} a_{11} & a_{12} & \cdots & a_{1n} \\ a_{21} & a_{22} & \cdots & a_{2n} \\ \vdots & \vdots & \ddots & \vdots \\ a_{m1} & a_{m1} & \cdots & a_{mn} \end{bmatrix}$$

Square Matrix: A matrix in which the number of columns is equal to the number of rows.

Addition and Subtraction of Matrices: To perform the addition or subtraction of matrices, the dimensions of two matrices must be equal:

$$\begin{bmatrix} a & b \\ c & d \end{bmatrix} + \begin{bmatrix} e & f \\ g & h \end{bmatrix} = \begin{bmatrix} a+e & b+f \\ c+g & d+h \end{bmatrix}$$

$$\begin{bmatrix} a & b \\ c & d \end{bmatrix} - \begin{bmatrix} e & f \\ g & h \end{bmatrix} = \begin{bmatrix} a-e & b-f \\ c-g & d-h \end{bmatrix}$$

Multiplication of two matrices: To perform the multiplication of two matrices, the matrices must be aligned, meaning that the number of columns in the first matrix is equal to the number of rows in the second matrix.

$$\underset{3\times4 \text{ matrix}}{\begin{bmatrix} \cdot & \cdot & \cdot & \cdot \\ \cdot & \cdot & \cdot & \cdot \\ 1 & 2 & 3 & 4 \end{bmatrix}} \underset{4\times5 \text{ matrix}}{\begin{bmatrix} \cdot & \cdot & \cdot & a & \cdot \\ \cdot & \cdot & \cdot & b & \cdot \\ \cdot & \cdot & \cdot & c & \cdot \\ \cdot & \cdot & \cdot & d & \cdot \end{bmatrix}} = \underset{3\times5 \text{ matrix}}{\begin{bmatrix} \cdot & \cdot & \cdot & \cdot & \cdot \\ \cdot & \cdot & \cdot & \cdot & \cdot \\ \cdot & \cdot & \cdot & x_{3,4} & \cdot \end{bmatrix}}$$

Matrix ID or Nature: One of the most important properties of a square matrix is its identity or nature, which is equal to the sum of the elements of the matrix and is represented by the symbol (χ).

❖ Note: The symmetry matrices are all squared and 3 * 3.

Matrix representation of symmetric groups E, i:

The symmetry function E is the only matrix whose all diagonal numbers are 1.

$\longrightarrow$ $\chi_E = 3$

$$\begin{bmatrix} -1 & 0 & 0 \\ 0 & -1 & 0 \\ 0 & 0 & -1 \end{bmatrix} \begin{bmatrix} x \\ y \\ z \end{bmatrix} = \begin{bmatrix} -x \\ -y \\ -z \end{bmatrix}$$

$\uparrow$
i operation

$\longrightarrow$ $\chi_i = -3$

$$\begin{bmatrix} 1 & 0 & 0 \\ 0 & 1 & 0 \\ 0 & 0 & 1 \end{bmatrix} \begin{bmatrix} x \\ y \\ z \end{bmatrix} = \begin{bmatrix} x \\ y \\ z \end{bmatrix}$$

$\uparrow$
E operation

Matrix representation of symmetry plates (σ): Coordinate axes create three coordinate plates; for example, if we consider the xy plane, the x and y coordinates on this plane remain constant while the coordinates of the z point change:

$$\sigma_{xy}: \begin{bmatrix} 1 & 0 & 0 \\ 0 & 1 & 0 \\ 0 & 0 & -1 \end{bmatrix} \begin{bmatrix} x \\ y \\ z \end{bmatrix} = \begin{bmatrix} x \\ y \\ -z \end{bmatrix} \qquad \sigma_{yz}: \begin{bmatrix} -1 & 0 & 0 \\ 0 & 1 & 0 \\ 0 & 0 & 1 \end{bmatrix} \begin{bmatrix} x \\ y \\ z \end{bmatrix} = \begin{bmatrix} -x \\ y \\ z \end{bmatrix}$$

σ operation

$$\sigma_{xz}: \begin{bmatrix} 1 & 0 & 0 \\ 0 & -1 & 0 \\ 0 & 0 & 1 \end{bmatrix} \begin{bmatrix} x \\ y \\ z \end{bmatrix} = \begin{bmatrix} x \\ -y \\ z \end{bmatrix}$$

Matrix representation of symmetric groups C_n , S_n:

$$\chi_{C_n} = 2\cos\theta + 1$$

C_n:
$$\begin{bmatrix} \cos\phi & \sin\phi & 0 \\ -\sin\phi & \cos\phi & 0 \\ 0 & 0 & 1 \end{bmatrix}\begin{bmatrix} x_1 \\ y_1 \\ z_1 \end{bmatrix} = \begin{bmatrix} x_2 \\ y_2 \\ z_2 \end{bmatrix}$$

C_n operation

$$\chi_{S_n} = 2\cos\theta - 1$$

S_n:
$$\begin{bmatrix} \cos\phi & \sin\phi & 0 \\ -\sin\phi & \cos\phi & 0 \\ 0 & 0 & -1 \end{bmatrix}\begin{bmatrix} x_1 \\ y_1 \\ z \end{bmatrix} = \begin{bmatrix} x_2 \\ y_2 \\ -z \end{bmatrix}$$

S_n operation ϕ is the angle of rotation

The result of multiplying two matrices, each representing a symmetric action, is a new matrix that represents a symmetric action itself.

Exercise: Prove $\sigma = S_1$ and $i = S_2$ using matrices

$$S_1 = C_1 . \, \sigma_h = E. \, \sigma_h = \sigma_h$$

$$\begin{bmatrix} 1 & 0 & 0 \\ 0 & 1 & 0 \\ 0 & 0 & 1 \end{bmatrix}\begin{bmatrix} 1 & 0 & 0 \\ 0 & 1 & 0 \\ 0 & 0 & -1 \end{bmatrix} = \begin{bmatrix} 1 & 0 & 0 \\ 0 & 1 & 0 \\ 0 & 0 & -1 \end{bmatrix} \qquad \rightarrow \qquad \chi_{S_1} = 1$$

$$S_2 = C_2(z). \, \sigma_{xy} = i$$

$$\begin{bmatrix} \cos 180 & -\sin 180 & 0 \\ \sin 180 & \cos 180 & 0 \\ 0 & 0 & 1 \end{bmatrix}\begin{bmatrix} 1 & 0 & 0 \\ 0 & 1 & 0 \\ 0 & 0 & -1 \end{bmatrix} = \begin{bmatrix} -1 & 0 & 0 \\ 0 & -1 & 0 \\ 0 & 0 & 1 \end{bmatrix}\begin{bmatrix} 1 & 0 & 0 \\ 0 & 1 & 0 \\ 0 & 0 & -1 \end{bmatrix}$$

$$= \begin{bmatrix} -1 & 0 & 0 \\ 0 & -1 & 0 \\ 0 & 0 & -1 \end{bmatrix}$$

Exercise: Show by multiplying matrices:

$$\sigma_{xz} \cdot \sigma_{yz} = C_2(z)$$

$$= \begin{bmatrix} 1 & 0 & 0 \\ 0 & -1 & 0 \\ 0 & 0 & 1 \end{bmatrix} \begin{bmatrix} -1 & 0 & 0 \\ 0 & 1 & 0 \\ 0 & 0 & 1 \end{bmatrix} = \begin{bmatrix} -1 & 0 & 0 \\ 0 & -1 & 0 \\ 0 & 0 & 1 \end{bmatrix}$$

Practice: Prove that in the C_{2v} point group the following relationships exist:

$$\sigma_{xz} \cdot C_2(z) = \sigma_{yz}$$

$$= \begin{bmatrix} 1 & 0 & 0 \\ 0 & -1 & 0 \\ 0 & 0 & 1 \end{bmatrix} \begin{bmatrix} -1 & 0 & 0 \\ 0 & -1 & 0 \\ 0 & 0 & 1 \end{bmatrix} = \begin{bmatrix} -1 & 0 & 0 \\ 0 & 1 & 0 \\ 0 & 0 & 1 \end{bmatrix}$$

$$\sigma_{yz} \cdot C_2(z) = \sigma_{xz}$$

$$= \begin{bmatrix} -1 & 0 & 0 \\ 0 & 1 & 0 \\ 0 & 0 & 1 \end{bmatrix} \begin{bmatrix} -1 & 0 & 0 \\ 0 & -1 & 0 \\ 0 & 0 & 1 \end{bmatrix} = \begin{bmatrix} 1 & 0 & 0 \\ 0 & -1 & 0 \\ 0 & 0 & 1 \end{bmatrix}$$

Exercise: Using the matrix representation, write the matrix of each symmetric element of the point C_{4v} with respect to the coordinates x, y, z:

$$C_{nv}: C_n \, , \, n\sigma_v$$

$$C_{4v} = C_4^1 \, , \, C_4^2 = C_2 \, , \, C_4^3 \, , \, C_4^4 = E \, , \, 4\sigma_v$$

3 * 3 square matrices

$$C_4^1(z) = \begin{bmatrix} \cos 90 & -\sin 90 & 0 \\ \sin 90 & \cos 90 & 0 \\ 0 & 0 & 1 \end{bmatrix} = \begin{bmatrix} 0 & -1 & 0 \\ 1 & 0 & 0 \\ 0 & 0 & 1 \end{bmatrix}$$

❖ If we separate the x and y axes from the z axis in the 3 * 3 matrix, two matrices will be obtained:

$$C_4(x, y) = \begin{bmatrix} 0 & -1 \\ 1 & 0 \end{bmatrix} \qquad \rightarrow \qquad \chi_{C_4(x,y)} = 0$$

$$C_4(z) = [1] \qquad \rightarrow \qquad \chi_{C_4(z)} = 1$$

As a result:

$$\chi_{C_2(x,y)} = -2, \ \chi_{C_2(z)} = 1 \quad C_2(z) = \begin{bmatrix} \cos 180 & -\sin 180 & 0 \\ \sin 180 & \cos 180 & 0 \\ 0 & 0 & 1 \end{bmatrix} = \begin{bmatrix} -1 & 0 & 0 \\ 0 & -1 & 0 \\ 0 & 0 & 1 \end{bmatrix}$$

$$\chi_{C_4^3(x,y)} = 0 \ , \quad C_4^3(z) = \begin{bmatrix} \cos 180 & -\sin 180 & 0 \\ \sin 180 & \cos 180 & 0 \\ 0 & 0 & 1 \end{bmatrix} = \begin{bmatrix} -1 & 0 & 0 \\ 0 & -1 & 0 \\ 0 & 0 & 1 \end{bmatrix}$$

$$\chi_{C_4^3(z)} = 1$$

$$\sigma_{xz} = \begin{bmatrix} 1 & 0 & 0 \\ 0 & -1 & 0 \\ 0 & 0 & 1 \end{bmatrix} \qquad\qquad \sigma_{yz} = \begin{bmatrix} -1 & 0 & 0 \\ 0 & 1 & 0 \\ 0 & 0 & 1 \end{bmatrix}$$

$$\sigma_d = \begin{bmatrix} 0 & 1 & 0 \\ 1 & 0 & 0 \\ 0 & 0 & 1 \end{bmatrix} \qquad\qquad \sigma_d{}' = \begin{bmatrix} 0 & -1 & 0 \\ -1 & 0 & 0 \\ 0 & 0 & 1 \end{bmatrix}$$

Irreducible View (Γ)

The following points must be considered in order to obtain irreducible representations in a character table:

1. The principle of representations: The condition of interactivity (perpendicular) is one of the basic conditions in the representation of groups; in other words, the sum of the productions of the representations in each other must be equal to zero.

2. The equation $\Sigma (\chi_R)^2 = h$ is true in the invariant representation.

3. The equation $\Sigma (\chi_E)^2 = h$ is true in the irreducible representation.

4. Each character table has a representation of total symmetry.

Exercise: Obtain the C_{2v} character table.

Solve: According to the fourth condition in the above condition, there is a total symmetry representation in each character table, so we put the number one in the first row of the invariant representation characters.

C_{2v}	E	$C_2(z)$	σ_{yz}	σ_{xz}
Γ_1	1	1	1	1

According to the third condition we write:

C_{2v}	E
Γ_1	1
Γ_2	X=1
Γ_3	Y=1
Γ_4	Z=1

$$x^2+y^2+z^2=4 \quad \rightarrow \quad x=y=z=1+1^{r} \rightarrow \quad \Sigma\,(\chi_E)^2=h$$

According to the first condition, the sum of the productions of the shows must be equal to zero. So if we consider two negatives, the result will be zero:

$$\Sigma\,\Gamma_1\Gamma_2=0 \rightarrow \quad 1\times 1 + 1\times A + 1\times B + 1\times C = 0$$

$$0 = 2 - 2 = (1-)\times 1 + (1-)\times 1 + (1+)\times 1 + 1\times 1 \quad \longrightarrow$$

C_{2v}	E	$C_2(z)$	σ_{yz}	σ_{xz}
Γ_1	1	1	1	1
Γ_2	1	A=1	B=-1	C=-1

According to the second condition we write:

$$\Sigma(\chi_R)^2=h \quad \rightarrow \quad 1^2 + 1^2 + (-1)^2 + (-1)^2 = 4$$

We will have:

C_{2v}	E	$C_2(z)$	σ_{xz}	σ_{yz}
Γ_1	1	1	1	1
Γ_2	1	1	−1	−1
Γ_3	1	−1	1	−1
Γ_4	1	−1	−1	1

Practice: Get irrepressible performances of the C_{3v} group.

Solve:

1. According to the representation of total symmetry, we set the number 1 for the first character of the invariant representation.

C_{3V}	E	$2C_3$	$3\sigma_v$
Γ_1	1	1	1

2. According to the third condition we write:

C_{3V}	E
Γ_1	1
Γ_2	X= 1
Γ_3	Y=1

$$x^2+y^2+z^2=6 \quad \rightarrow \quad x =1 \quad , \quad y=2+١٢ \rightarrow \quad \Sigma\,(\chi_E)^2=h$$

3. According to the first condition, the sum of the product of the representations in each other must be equal to zero (but since the values of x and y are different, the first condition must be calculated for each row separately):

$$\Sigma\,\Gamma_1\Gamma_2.\ x =0 \rightarrow \quad 1\times 1\times 1 + a\times 1\times 2 + b\times 1\times 3= 0 \quad \rightarrow \quad a =1, b=-1$$

$$\Sigma\,\Gamma_1\Gamma_3.\ x =0 \rightarrow \quad 1\times 1\times 2 + 2\times 1\times (a') + 3\times 1\times (b')= 0$$

$$\rightarrow \quad a' =-1, b'=0$$

To be sure, if we multiply both different rows of the irreducible representation and affect the coefficients obtained, the result should be zero:

$$\Sigma\,\Gamma_2\Gamma_3.\ x =0 \rightarrow \quad 1\times 2\times 1 + 1\times (-1)\times 2 + (-1)\times 0\times 3= 0$$

4. According to the second condition we write:

$$\Sigma(\chi_R)^2=h \quad \rightarrow \quad 1^2 + 2\times 1^2 + 3\times (-1)^2 =6 \ :\text{for the second line}$$

$$\Sigma(\chi_R)^2 = h \quad \rightarrow \quad 2^2 + 2 \times (-1)^2 + 3 \times 0^2 = 6 \text{ :for the third line}$$

We will have:

C_{3V}	E	$2C_3$	$3\sigma_v$
Γ_1	1	1	1
Γ_2	1	1	-1
Γ_3	1	-1	0

The symbols A, B, two-dimensional E and three-dimensional T are used to name one-dimensional irreducible representations. If one-dimensional representation is symmetric with respect to the axis of symmetry, the symbol A is used and if not symmetric with respect to the axis of symmetry, the symbol B is used.

❖ **Note:** In these representations, if the (+1) number indicates a symmetry, and if the (-1) number indicates an asymmetry.

Index (1) is used for a display that is symmetrical to the home screen and index (2) is used for a display that is asymmetric to the home screen.

If present in molecule σ_h, symmetric representation of σ_h uses the sign (′) and asymmetric representation of σ_h uses the sign (″).

Convert Reducible View to Irreducible View

The following equation is used to determine what an irreducible representation consists of a reducible representation; in other words, the following relation is used to determine each irreducible representation repeatedly in the desired reducible representation:

$$n\Gamma = \frac{1}{h} \Sigma_g\, n_g \cdot \chi_R \cdot \chi_\Gamma$$

h = group rank

n_g = The number of symmetric operations in the class g

χ_R = Decrease id display

χ_Γ = Irreversibly reduced id representation

Exercise: Obtain irreversible performances of the C_{3v} group from the reduced performances of this group.

$$n\Gamma = \frac{1}{h} \sum_g n_g \cdot \chi_R \cdot \chi_\Gamma$$

$\{\Gamma_a\}$

$$n_{A1} = n\Gamma_1 = \frac{1}{6}[1 \times 1 \times 5 + 2 \times 1 \times 2 + 3 \times 1 \times (-1)] = 1$$

$$n_{A2} = n\Gamma_2 = \frac{1}{6}[1 \times 1 \times 5 + 2 \times 1 \times 2 + 3 \times (-1) \times (-1)] = 2$$

$$n_E = n\Gamma_3 = \frac{1}{6}[1 \times 2 \times 5 + 2 \times (-1) \times 2 + 3 \times 0 \times (-1)] = 1$$

$$\Gamma_a = \Gamma_1 + 2\Gamma_2 + \Gamma_3 \quad \rightarrow \quad \Gamma_a = A_1 + 2A_2 + E$$

$\{\Gamma b\}$

$$n_{A1} = n\Gamma_1 = \frac{1}{6}[1 \times 1 \times 7 + 2 \times 1 \times 1 + 3 \times 1 \times (-3)] = 0$$

$$n_{A2} = n\Gamma_2 = \frac{1}{6}[1 \times 1 \times 7 + 2 \times 1 \times 1 + 3 \times (-1) \times (-3)] = 3$$

$$n_E = n\Gamma_3 = \frac{1}{6}[1 \times 2 \times 7 + 2 \times (-1) \times 1 + 3 \times 0 \times (-3)] = 2$$

$$\Gamma_b = 3\Gamma_2 + 2\Gamma_3 \quad \rightarrow \quad \Gamma_b = 3A_2 + 2E$$

C_{3V}	E	$2C_3$	$3\sigma_v$
$A_1 = \Gamma_1$	1	1	1
$A_2 = \Gamma_2$	1	1	−1
$E = \Gamma_3$	2	−1	0
Γ_a	5	2	-1
Γ_b	7	1	-3

Character Table (Identification Table)

We use tables called character tables to apply group theory and molecular symmetry. Here's a table for the C_{3v} symmetry group:

C_{3v}	E	$2C_3$	$3\sigma_v$	Linear bases	Secondary Exponentiation bases	Third Exponentiation bases
A_1	1	1	1	Z	$X^2+Y^2+Z^2$	Z^3, $X(X^2-3Y^2)$
A_2	1	1	−1			
E	2	−1	0	R_z		$Y(3X^2-Y^2)$
				$(X,Y)(R_x , R_y)$	(X^2-Y^2, XY) (XZ, YZ)	(XZ^2, YZ^2)

Table description

> The first column contains the symbols for each irreducible representation in the group. The number obtained at the intersection of a row and a column is an intractable character for the class in question.

> The characters in the column below the same element represent the dimensions of the display.

Note: For groups such as C_{2h} D_{2h} D_{4h} O_h which have the center of inversion, the subgroup g$'$ uses the symmetric meaning of the center and u^r the symmetric.

> Linear Bases Columns: In this column, the symbol R_x, R_y and R_z means rotation

---

around the x, y, z axes, and the symbols x, y, z represent the orbitals of P_x, P_y, P_z, as well as the transitions along the mentioned axes.

➤ Secondary Exponentiation Bases column: This orbital column is displayed in different ways:

➤ Orbital S $\longrightarrow$ $x^2 + y^2 + z^2$ or $x^2 + y^2$ or k^2 و y^2 و z^2

➤ Orbital d_{z^2} $\longrightarrow$ z^2 or $z^2 - x^2 - y^2$ 2

➤ Orbital d_{xz} و d_{yz} و d_{xy} $\longrightarrow$ xy , xz , yz

➤ Third Exponentiation Bases Column: This column represents the F orbitals.

Exercise: The P_x, P_y, P_z orbitals in the C_{2v} group belong to which irreducible representations?

C_{2v}	E	$C_2(z)$	σ_{xz}	σ_{yz}
A_1	1	1	1	1
A_2	1	1	−1	−1
B_1	1	−1	1	−1
B_2	1	−1	−1	1

Solve: We first examine the p_x orbital and examine the effect of each element on this orbital:

From the numbers obtained, it can be seen that the P_x orbital belongs to row B_1 of the table:

Then we examine the p_z orbital and plot the effect of each element on this orbital:

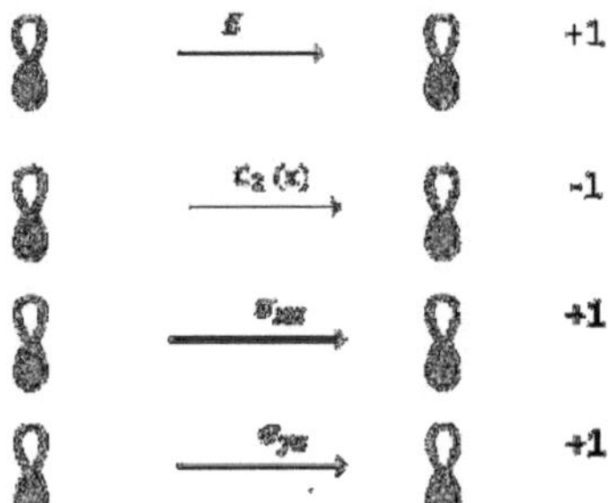

From the numbers obtained, it can be seen that P_z orbital belongs to row A1 of the table:

$$P_z \rightarrow A_1$$

Finally, consider the P_y orbital:

From the numbers obtained, it can be seen that the P_y orbital belongs to row B_2 of the table:

$$P_y \rightarrow B_2$$

Chapter III

Exercise

Multiplicities in the character table

The product of two invariant representations in the character table is a representation of the same group, although these numbers may sometimes not appear in the table, which would have to be reduced.

Rules for Multiplication Evaluation of Intractable Views in the Character Table

A ×A = A	B ×A = B	E×A = E	T ×A = T
A ×B = B	B ×B = A	E ×B = E	T ×B = T
A ×E = E	B ×E = E	❋ = E ×E	❋ = T ×E
A ×T = T	B ×T = T	❋ = E ×T	❋ = T ×T
	g ×g = g	' ×' = '	1 = 1×1
	g ×u = u	' ×" = "	2 = 2× 1
	u ×g = u	" ×' = "	2 = 1× 2
	u ×u = g	" ×" = '	1 = 2× 2

❋This amount must be calculated

Exercise 1: Obtain the results of each of the following products from table C_{3v}:

C_{3V}	E	$2C_3$	$3\sigma_v$
A_1	1	1	1
A_2	1	1	−1
E	2	−1	0
$A_1 \times A_2$	1	1	-1
$A_1 \times E$	1	1	1
$A_2 \times E$	1	1	-1
$E \times E$	4	1	0

$$\rightarrow A_1 \times A_2 = A_2$$
$$A_1 \times E = E$$
$$A_2 \times E = E$$

The product of $E \times E$ representations are reducible and must be reduced according to the formula and become irreducible:

$$n_\Gamma = \frac{1}{h} \sum_g n_g \cdot \chi_R \cdot \chi_\Gamma$$

$$n_{A1} = \frac{1}{6}[1 \times 1 \times 4 + 2 \times 1 \times 1 + 3 \times 1 \times 0] = 1$$

$$n_{A2} = \frac{1}{6}[1 \times 1 \times 4 + 2 \times 1 \times 1 + 3 \times (-1) \times 0] = 1$$

$$n_E = \frac{1}{6}[1 \times 2 \times 4 + 2 \times (-1) \times 1 + 3 \times 0 \times 0] = 1$$

$$\rightarrow E \times E = A_1 + A_2 + E$$

Exercise 2: Obtain the results of each of the following products from the C_{4v} table:

C_{4V}	E	C_2	$2C_4$	$2\sigma_v$	$2\sigma_d$
A_1	1	1	1	1	1
A_2	1	1	1	-1	-1
B_1	1	1	-1	1	-1
B_2	1	1	-1	-1	1
E	2	-2	0	0	0
$A_1 \times E$	2	-2	0	0	0
$E \times E$	4	4	0	0	0
$B_1 \times E$	2	-2	0	0	0
$A_1 \times B_2$	1	1	-1	-1	1
$B_2 \times B_2$	1	1	1	1	1

The product of E × E representations are reducible and must be reduced according to the formula and become irreducible:

$$n\Gamma = \frac{1}{h}\Sigma_g\, n_g \cdot \chi_R \cdot \chi_\Gamma$$

$$n_{A1} = \frac{1}{8}[1\times1\times4 + 1\times1\times4 + 2\times1\times0 + 2\times1\times0 + 2\times1\times0] = 1$$

$$n_{A2} = \frac{1}{8}[1\times1\times4 + 1\times1\times4 + 0 + 0 + 0] = 1$$

$$n_{B1} = \frac{1}{8}[1\times1\times4 + 1\times1\times4 + 0 + 0 + 0] = 1$$

$$n_{B2} = \frac{1}{8}[1\times1\times4 + 1\times1\times4 + 0 + 0 + 0] = 1$$

$$n_{B1} = \frac{1}{8}[1\times1\times4 + 1\times(-2)\times4 + 0 + 0 + 0] = 0$$

$$\rightarrow E \times E = A_1 + A_2 + B_1 + B_2$$

Exercise 3. In point group T_d, what are the indivisible productions of the productions of

- ✓ $T_1 \times T_2$,
- ✓ $T_1 \times T_1$,
- ✓ $T_2 \times T_2$

T_d	E	$8C_3$	$3C_2$	$6S_4$	$6\sigma_d$
A_1	1	1	1	1	1
A_2	1	1	1	-1	-1
E	2	-1	2	0	0
T_1	3	0	-1	1	-1
T_2	3	0	-1	-1	1
$T_1 \times T_1$	9	0	1	-1	-1
$T_1 \times T_2$	9	0	1	1	1
$T_2 \times T_2$	9	0	1	1	1

First, we obtain the product of $T_1 \times T_2$ representations:

$$n\Gamma = \frac{1}{h}\sum_g n_g \cdot \chi_R \cdot \chi_\Gamma$$

$$n_{A1} = \frac{1}{24}[1\times1\times9 + 8\times1\times0 + 3\times1\times1 + 6\times1\times(-1) + 6\times1\times(-1)] = 0$$

$$n_{A2} = \frac{1}{24}[1\times1\times9 + 8\times1\times0 + 3\times1\times1 + 6\times(-1)\times(-1) + 6\times(-1)\times(-1)] = 1$$

$$n_{E} = \frac{1}{24}[1\times2\times9 + 8\times(-1)\times0 + 3\times2\times1 + 6\times0\times(-1) + 6\times0\times(-1)] = 1$$

$$n_{T1} = \frac{1}{24}[1\times3\times9 + 8\times0\times0 + 3\times(-1)\times1 + 6\times1\times(-1) + 6\times(-1)\times(-1)] = 1$$

$$n_{T2} = \frac{1}{24}[1\times3\times9 + 8\times0\times0 + 3\times(-1)\times1 + 6\times(-1)\times(-1) + 6\times1\times(-1)] = 1$$

$\rightarrow T_1 \times T_2 = A_2 + E + T_1 + T_2$

Now we obtain the product of $T_1 \times T_2$ representations:

$$n\Gamma = \frac{1}{h} \Sigma_g \, n_g \cdot \chi_R \cdot \chi_\Gamma$$

$$n_{A1} = \frac{1}{24}[1 \times 1 \times 9 + 8 \times 1 \times 0 + 3 \times 1 \times 1 + 6 \times 1 \times 1 + 6 \times 1 \times 1] = 1$$

$$n_{A2} = \frac{1}{24}[1 \times 1 \times 9 + 8 \times 1 \times 0 + 3 \times 1 \times 1 + 6 \times (-1) \times 1 + 6 \times (-1) \times 1] = 0$$

$$n_E = \frac{1}{24}[1 \times 2 \times 9 + 8 \times (-1) \times 0 + 3 \times 2 \times 1 + 6 \times 0 \times 1 + 6 \times 0 \times 1] = 1$$

$$n_{T1} = \frac{1}{24}[1 \times 3 \times 9 + 8 \times 0 \times 0 + 3 \times (-1) \times 1 + 6 \times 1 \times 1 + 6 \times (-1) \times 1] = 1$$

$$n_{T2} = \frac{1}{24}[1 \times 3 \times 9 + 8 \times 0 \times 0 + 3 \times (-1) \times 1 + 6 \times (-1) \times 1 + 6 \times 1 \times 1] = 1$$

$\rightarrow T_1 \times T_1 = T_2 \times T_2 = A_1 + E + T_1 + T_2$

Exercise 4: Investigate Are the two symmetry elements C_3 in the point group C_{3v} in the same class?

$$C_{3v}= E \quad 2C_3 \quad 3\sigma_v$$

The condition of being on a floor is to be conjugate:

$$\sigma_v \times C_3^1 \times \sigma_v^{-1} = C_3^2$$

Explanation of the above figure: In the first step we affect σ_v by moving 2 and 3, then we affect C_3^1, which will cause 120° to move numbers, and in the third step we affect σ_v^{-1} by 1 and 3. They will move. Finally, if we again affect C_3^1, the initial form will emerge, and we can conclude that the two symmetric elements C_3^1 and C_3^2 are in the same class;

$$C_3^1 \times \sigma_v \times C_3^{-1} = \sigma_v''$$

$$C_3^2 \times \sigma_v \times C_3^{-2} = \sigma_v'$$

So the three symmetric elements σ_v, σ_v' and σ_v'' are in the same class.

Exercise 5: In the D_{3d} group write a matrix based on x, y, z and reduce it to irreducible representations.

D_{3d} (3m)	E	$2C_3$	$3C_2$	i	$2S_6$	$3\sigma_d$	$h = 12$	
A_{1g}	1	1	1	1	1	1		$x^2 + y^2, z^2$
A_{2g}	1	1	−1	1	1	−1	R_z	
E_g	2	−1	0	2	−1	0	(R_x, R_y)	$(x^2 - y^2, xy)\,(zx, yz)$
A_{1u}	1	1	1	−1	−1	−1		
A_{2u}	1	1	−1	−1	−1	1	z	
E_u	2	−1	0	−2	1	0	(x, y)	

Γ_{xyz}	3	0	-1	-3	0	1

$$E=\begin{bmatrix}1 & 0 & 0\\0 & 1 & 0\\0 & 0 & 1\end{bmatrix} \rightarrow \chi_E=3 \qquad i=\begin{bmatrix}-1 & 0 & 0\\0 & -1 & 0\\0 & 0 & -1\end{bmatrix} \rightarrow \chi_i=-3$$

$$\sigma_d=\begin{bmatrix}-1 & 0 & 0\\0 & 1 & 0\\0 & 0 & 1\end{bmatrix} \rightarrow \chi_{\sigma d}=1 \qquad C_3=\begin{bmatrix}\cos120 & -\sin120 & 0\\\sin120 & \cos120 & 0\\0 & 0 & 1\end{bmatrix} \rightarrow \chi_{C3}=2\cos120+1=0$$

$$C_2=\begin{bmatrix}\cos180 & -\sin180 & 0\\\sin180 & \cos180 & 0\\0 & 0 & 1\end{bmatrix} \rightarrow \chi_{C2}=2\cos180+1=-1$$

$$S_6=\begin{bmatrix}\cos60 & -\sin60 & 0\\\sin60 & \cos60 & 0\\0 & 0 & -1\end{bmatrix} \rightarrow \chi_{S6}=2\cos60-1=0$$

$$n\Gamma = \frac{1}{h}\sum_g n_g \cdot \chi_R \cdot \chi_\Gamma$$

$$n_{A1g}=\frac{1}{12}[1\times1\times3+2\times1\times0+3\times1\times(-1)+1\times1\times(-3)+2\times1\times0+3\times1\times1]=0$$

$$n_{A2g}=\frac{1}{12}[1\times1\times3+2\times1\times0+3(-1)\times(-1)+1\times1\times(-3)+2\times1\times0+3\times(-1)\times1]=0$$

$$n_{Eg}=\frac{1}{12}[1\times2\times3+2\times(-1)\times0+3\times0\times(-1)+1\times2\times(-3)+2\times(-1)\times0+3\times0\times1]=0$$

$$n_{A1u}=\frac{1}{12}[1\times1\times3+0+3\times1\times(-1)+1\times(-1)\times(-3)+0+3\times(-1)\times1]=0$$

$$n_{A2u}=\frac{1}{12}[1\times1\times3+0+3\times(-1)\times(-1)+1\times(-1)\times(-3)+0+3\times1\times1]=1$$

$$n_{Eu}=\frac{1}{12}[1\times2\times3+0+0+1\times(-2)\times(-3)+0+0]=1$$

$$\rightarrow \Gamma xyz = A2u + Eu$$

Chapter IV

Application of group theory

Application of Group Theory

Conventional vibration modes: Conventional modes of vibration are vibrational modes that lead to absorption in IR. Because each atom has three directions of motion along the x, y, and z axes, then each atom will have three degrees of freedom.

There is a total of 3N degrees of freedom for each particle, not all of these degrees of vibration are related to vibrational motions.

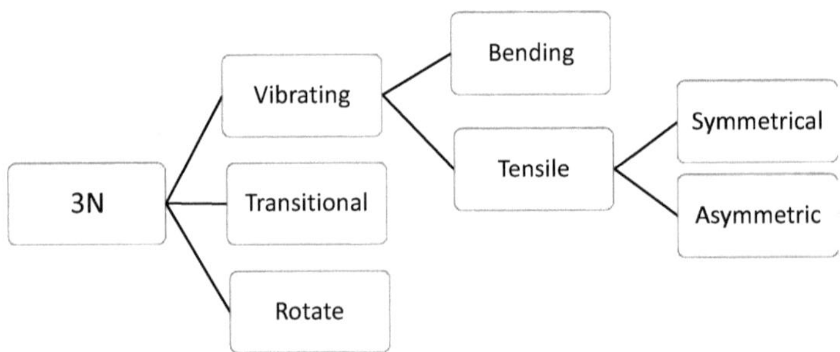

Conventional modes of vibration

In linear molecules = 3N – 5

In nonlinear molecules = 3N – 6

Explain that, whether linear or nonlinear, we have three degrees of transient freedom; there are three rotational motions per nonlinear molecule, and so the number of vibrational motions $3N\text{-}6$ will be in the nonlinear molecule. While in a linear molecule since one of the coordinate axes is also the axis of the molecule, there is no movement of atoms on this axis, so the number of vibrational motions in 3N-5 will be the same.

IR spectrometers record vibrational patterns between 400 and 4000 nm. In the discussion of vibrational modes, polarization can also be used in the Raman spectrometer rather than the change in the dipole moment in the IR spectrometer.

In molecules having center of symmetry, vibrations that are active in Raman are inactive in IR. It should be noted that in Raman, ultraviolet (UV) light is used, while in IR, infrared light is used.

There are generally three types of IR spectroscopy that measure these distances relative to the UV region:

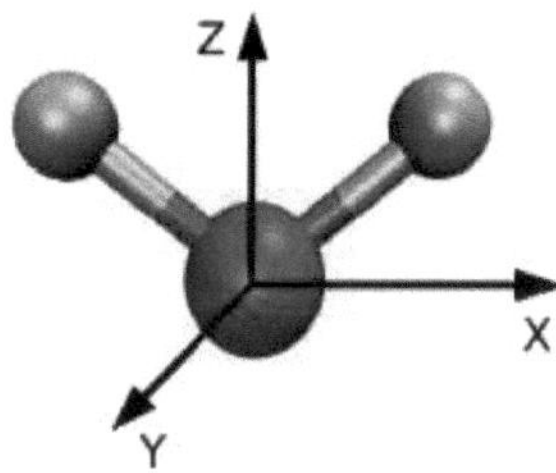

The observed frequency for each vibration is obtained from:

$$\upsilon = \frac{1}{2\pi}\sqrt{\frac{k}{\mu}}$$

k= spring constant

$$\text{Reduced mass} = \mu = \frac{m1\,m2}{m1+m2}$$

Exercise 1: Find the vibrational symmetry of the water molecule.

C_{2v}: E, $C_2(z)$, σ_{xz}, σ_{yz}

We need to specify the representations of the molecule, first we need to obtain the matrices used in the molecule and then the character of these matrices. By specifying the corresponding character, a totally reducible representation is obtained that divides this representation into irreducible representations, and finally, by removing rotational and transient modes, we obtain vibrational modes:

E	x_1	y_1	z_1	x_2	y_2	z_2	x_3	y_3	z_3
X_1	1	0	0	0	0	0	0	0	0
Y_1	0	1	0	0	0	0	0	0	0
Z_1	0	0	0	0	0	0	0	0	0
X_2	0	0	0	1	0	0	0	0	0
Y_2	0	0	0	0	1	0	0	0	0
Z_2	0	0	0	0	0	1	0	0	0
X_3	0	0	0	0	0	0	1	0	0
Y_3	0	0	0	0	0	0	0	1	0
Z_3	0	0	0	0	0	0	0	0	1

$$\chi_E = 9$$

$$E = \begin{vmatrix} \text{Cos}360 & -\text{Sin}360 & 0 \\ \text{Sin}360 & \text{Cos}360 & 0 \\ 0 & 0 & 1 \end{vmatrix} \rightarrow \begin{vmatrix} 1 & 0 & 0 \\ 0 & 1 & 0 \\ 0 & 0 & 1 \end{vmatrix} \rightarrow \chi_E = 3$$

$C_2(z)$	x_1	y_1	z_1	x_2	y_2	z_2	x_3	y_3	z_3
X_1	0	0	0	0	0	0	-1	0	0
Y_1	0	0	0	0	0	0	0	-1	0
Z_1	0	0	0	0	0	0	0	0	1
X_2	0	0	0	-1	0	0	0	0	0
Y_2	0	0	0	0	-1	0	0	0	0
Z_2	0	0	0	0	0	1	0	0	0
X_3	-1	0	0	0	0	0	0	0	0
Y_3	0	-1	0	0	0	0	0	0	0
Z_3	0	0	1	0	0	0	0	0	0

$$C_2(z) = \begin{vmatrix} \text{Cos}180 & -\text{Sin}180 & 0 \\ \text{Sin}180 & \text{Cos}180 & 0 \\ 0 & 0 & 1 \end{vmatrix} \rightarrow \begin{vmatrix} -1 & 0 & 0 \\ 0 & -1 & 0 \\ 0 & 0 & 1 \end{vmatrix} \rightarrow \chi_{C2(z)} = -1$$

σ_{XZ}	x_1	y_1	z_1	x_2	y_2	z_2	x_3	y_3	z_3
X_1	1	0	0	0	0	0	0	0	0
Y_1	0	-1	0	0	0	0	0	0	0
Z_1	0	0	1	0	0	0	0	0	0
X_2	0	0	0	1	0	0	0	0	0
Y_2	0	0	0	0	-1	0	0	0	0
Z_2	0	0	0	0	0	1	0	0	0
X_3	0	0	0	0	0	0	1	0	0
Y_3	0	0	0	0	0	0	0	-1	0
Z_3	0	0	0	0	0	0	0	0	1

$$\sigma_{XZ} = \begin{vmatrix} -1 & 0 & 0 \\ 0 & 1 & 0 \\ 0 & 0 & 1 \end{vmatrix} \rightarrow \chi_{\sigma xz} = 1$$

σ_{YZ}	x_1	y_1	z_1	x_2	y_2	z_2	x_3	y_3	z_3
X_1	0	0	0	0	0	0	-1	0	0
Y_1	0	0	0	0	0	0	0	1	0
Z_1	0	0	0	0	0	0	0	0	1
X_2	0	0	0	-1	0	0	0	0	0
Y_2	0	0	0	0	1	0	0	0	0
Z_2	0	0	0	0	0	1	0	0	0
X_3	-1	0	0	0	0	0	0	0	0
Y_3	0	1	0	0	0	0	0	0	0
Z_3	0	0	1	0	0	0	0	0	0

$$\sigma_{yz} = \begin{vmatrix} -1 & 0 & 0 \\ 0 & 1 & 0 \\ 0 & 0 & 1 \end{vmatrix} \rightarrow \chi_{\sigma yz} = 1$$

Now we have to write the C_{2v} group invariant representation table:

C_{2v}	E	$C_2(z)$	σ_{xz}	σ_{yz}	Linear bases
A_1	1	1	1	1	Z
A_2	1	1	-1	-1	R_z
B_1	1	-1	1	-1	X, R_y
B_2	1	-1	-1	1	Y, R_x
Γ_{xyz}	9	-1	1	3	

$$n_\Gamma = \frac{1}{h} \sum_g n_g \cdot \chi_R \cdot \chi_\Gamma$$

$$n_{A1} = \frac{1}{4}[1 \times 1 \times 9 + 1 \times 1 \times (-1) + 1 \times 1 \times 1 + 1 \times 1 \times 3] = 3$$

$$n_{A2} = \frac{1}{4}[1 \times 1 \times 9 + 1 \times 1 \times (-1) + 1 \times (-1) \times 1 + 1 \times (-1) \times 3] = 1$$

$$n_{B1} = \frac{1}{4}[1 \times 1 \times 9 + 1 \times (-1) \times (-1) + 1 \times 1 \times 1 + 1 \times (-1) \times 3] = 2$$

$$n_{B2} = \frac{1}{4}[1 \times 1 \times 9 + 1 \times (-1) \times (-1) + 1 \times (-1) \times 1 + 1 \times 1 \times 3] = 3$$

$$\rightarrow \Gamma_{xyz} = 3A_1 + A_2 + 2B_1 + 3B_2$$

The representations in the columns written in x, y, and z indicate that the transition gamma is in the same direction. So in the above character table, transitive gamma is:

$\Gamma_{\text{Transitional}} = A_1 , B_1 , B_2$

R_x, R_y, R_z representations of rotational gamma:

$\Gamma_{\text{ROTATE}} = A2 , B1 , B2$

To obtain the vibrational gamma, we can:

$$\Gamma_{\textbf{Vibratting}} = \Gamma_{\textbf{xyz}} - [\Gamma_{\textbf{Transitional}} + \Gamma_{\textbf{Rotate}}]$$

$$\Gamma_{\textbf{Vibrating}} = 3A_1 + A_2 + 2B_1 + 3B_2 - [A_1 + A_2 + 2B_1 + 2B_2] = 2A_1 + B_2$$

In symmetric vibration, one of the sentences A_1 can be a component of symmetric tensile vibration and the other A_1 is related to symmetric flexural vibration, and finally B_2 refers to asymmetric tensile vibration.

The following can be used to determine conventional vibration modes:

These methods can be represented as vectors along the angles and lengths of the links. If they are moved after symmetry, then we consider them zero, and if not, they will be +1:

C_{2V}	E	$C_2(z)$	σ_{xz}	σ_{yz}
$\Gamma_{\textbf{inner}}$	3	1	1	3

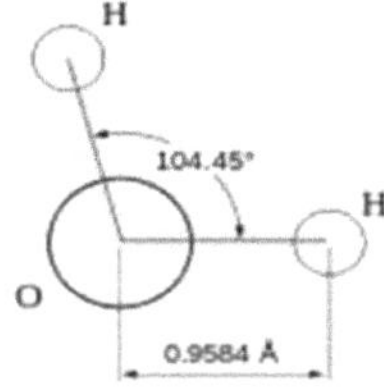

To explain that:

- ✓ None of the inner coordinates (α, r_1, r_2) change due to applying the same symmetry element (E), so we write 3 in the column below E.
- ✓ Due to the application of the symmetry element C_2 (z), the angle (α) does not change, so in the column below we write the +1 symmetry element.
- ✓ Due to the application of the symmetry element σ_{xz}, the angle (α) does not change, so in the column below we write the +1 symmetry element.

✓ None of the inner coordinates (α, r_1, r_2) of the symmetry element σ_{xz} change, so in the column below we write the symmetry of the number 3.

Once again, we write the character table of the invariant representation of C_2 (z) and compute it with its internal coordinates:

C_{2v}	E	$C_2(z)$	σ_{xz}	σ_{yz}
A_1	1	1	1	1
A_2	1	1	-1	-1
B_1	1	-1	1	-1
B_2	1	-1	-1	1
Γ_{inner}	3	1	1	3

$$n\Gamma = \frac{1}{h}\sum_g n_g \cdot \chi_R \cdot \chi_\Gamma$$

$$n_{A1} = \frac{1}{4}[1\times1\times3 + 1\times1\times1 + 1\times1\times1 + 1\times1\times3] = 2$$

$$n_{A2} = \frac{1}{4}[1\times1\times3 + 1\times1\times1 + 1\times(-1)\times1 + 1\times(-1)\times3\,] = 0$$

$$n_{B1} = \frac{1}{4}[1\times1\times3 + 1\times(-1)\times1 + 11\times1 + 1\times(-1)\times3] = 0$$

$$n_{B2} = \frac{1}{4}[1\times1\times3 + 1\times(-1)\times\,1 + 1\times(-1)\times1 + 1\times1\times3] = 1$$

$$\rightarrow \Gamma_{inner} = 2A_1 + B_2$$

Now we must obtain the sum of the two rows Γ_r and Γ_α such that the sum of these two times the interior of Γ_{inner} is:

C_{2v}	E	$C_2(z)$	σ_{xz}	σ_{yz}
Γ_{inner}	3	1	1	3
Γ_r	2	0	0	2
Γ_α	1	1	1	1

Once again, we write the C_{2v} point group character table, instead of the Γ_{inner}, we write the two characters Gr and $\Gamma\alpha$, and finally calculate Γ_r and $\Gamma\alpha$:

C_{2v}	E	$C_2(z)$	σ_{xz}	σ_{yz}
A_1	1	1	1	1
A_2	1	1	-1	-1
B_1	1	-1	1	-1
B_2	1	-1	-1	1
Γ_r	2	0	0	2
Γ_α	1	1	1	1

$$n\Gamma = \frac{1}{h}\sum_g n_g \cdot \chi_R \cdot \chi_\Gamma$$

$$n_{A1} = \frac{1}{4}[1 \times 1 \times 2 + 1 \times 1 \times 0 + 1 \times 1 \times 0 + 1 \times 1 \times 2] = 1$$

$$n_{A2} = \frac{1}{4}[1 \times 1 \times 2 + 1 \times 1 \times 0 + 1 \times (-1) \times 0 + 1 \times (-1) \times 2] = 0$$

$$n_{B1} = \frac{1}{4}[1 \times 1 \times 2 + 1 \times (-1) \times 0 + 1 \times 1 \times 0 + 1 \times (-1) \times 2] = 0$$

$$n_{B2} = \frac{1}{4}[1 \times 1 \times 2 + 1 \times (-1) \times 0 + 1 \times (-1) \times 0 + 1 \times 1 \times 2] = 1$$

$$\rightarrow \Gamma_R = A_1 + B_2$$

$$n\Gamma = \frac{1}{h}\sum_g n_g \cdot \chi_R \cdot \chi_\Gamma$$

$$n_{A1} = \frac{1}{4}[1 \times 1 \times 1 + 1 \times 1 \times 1 + 1 \times 1 \times 1 + 1 \times 1 \times 1] = 1$$

$$n_{A2} = \frac{1}{4}[1 \times 1 \times 1 + 1 \times 1 \times 1 + 1 \times (-1) \times 1 + 1 \times (-1) \times 1] = 0$$

$$n_{B1} = \frac{1}{4}[1 \times 1 \times 1 + 1 \times (-1) \times 1 + 1 \times 1 \times 1 + 1 \times (-1) \times 1] = 0$$

$$n_{B2} = \frac{1}{4}[1 \times 1 \times 1 + 1 \times (-1) \times 1 + 1 \times (-1) \times 1 + 1 \times 1 \times 1] = 0$$

$$\rightarrow \Gamma_\alpha = A_1$$

When we deal with complex molecules, the said methods will not work.

The atoms that are effective for determining the Cartesian gamma Γ_{xyz} are atoms that cannot be displaced by symmetric action; thus, the number of non-displaced atoms as a result of symmetric action can be determined and then the character of the matrix for each symmetric action can be determined. Now we can apply this method to the water molecule:

C_{2v}	E	$C_2(z)$	σ_{xz}	σ_{yz}
A_1	1	1	1	1
A_2	1	1	-1	-1
B_1	1	-1	1	-1
B_2	1	-1	-1	1
Γ_{xyz}	9	-1	1	3

We already had

$$E=\begin{bmatrix}1 & 0 & 0\\ 0 & 1 & 0\\ 0 & 0 & 1\end{bmatrix} \rightarrow \chi_E=3 \qquad\qquad C_2(z)=\begin{bmatrix}-1 & 0 & 0\\ 0 & -1 & 0\\ 0 & 0 & 1\end{bmatrix} \rightarrow \chi_{C3(z)}=-1$$

$$\sigma_{xz}=\begin{bmatrix}1 & 0 & 0\\ 0 & -1 & 0\\ 0 & 0 & 1\end{bmatrix} \rightarrow \chi_{\sigma xz}=1 \qquad\qquad \sigma_{yz}=\begin{bmatrix}-1 & 0 & 0\\ 0 & 1 & 0\\ 0 & 0 & 1\end{bmatrix} \rightarrow \chi_{\sigma yz}=1$$

So we can write:

ymmetrical Elements	χ	The number of atoms not displaced	Γ_{xyz}
E	3	3	9
$C_2(Z)$	-1	1	-1
σ_{XZ}	1	1	1
σ_{YZ}	1	3	3

In this table, we substitute the number 1 for every atom not moved. Therefore, to obtain a Cartesian gamma, we must multiply the number of atoms displaced as a result of applying the symmetric element to the matrix character corresponding to that symmetric element:

$$\Gamma_{xyz}= \chi \times \text{The number of atoms not displaced}$$

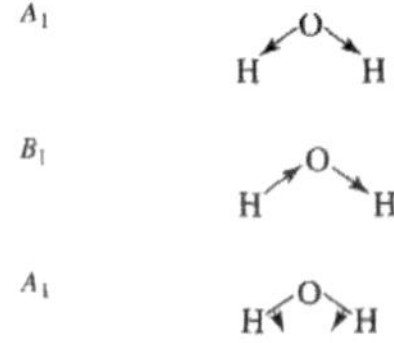

Figure19. The vibrational Modes of Water

Exercise 2: Calculate the Cartesian gamma for the CO_3^{-2} molecule.

$$D_{3h}$$

Solve: We already had:

$$\chi_{Cn} = 2\cos\theta + 1$$

$$\chi_{Sn} = 2\cos\theta - 1$$

$$\chi_E = +3$$

$$\chi_\sigma = +1$$

So:

Symmetrical Elements	χ	The number of atoms not displaced	Γ_{xyz}
E	3	4	12
C_3	0	1	0
C_2	-1	2	-2
S_3	1	2	2
σ_v	-2	1	-2
σ_h	1	4	4

Then we will have:

D_{3h}	E	$2C_3$	$3C_2$	σ_h	$2S_3$	$3\sigma_v$	Linear bases
A_1'	1	1	1	1	1	1	
A_2'	1	1	-1	1	1	-1	R_z
E'	2	-1	0	2	-1	0	(x, y)
A_1''	1	1	1	-1	-1	-1	

A_2''	1	1	-1	-1	-1	1	Z
E''	2	-1	0	-2	1	0	(R_x , R_y)
Γ_{XYZ}	12	0	-2	4	-2	2	
Γ_r	3	0	1	3	0	1	
Γ_α	2	0	1	3	0	1	

$\Gamma_{Transitional} = E' + A_2''$

$\Gamma_{Rotate} = A_2' + E''$

Now we reduce the $\Gamma_{Cartesian}$:

$$n\Gamma = \frac{1}{h}\sum_g n_g \cdot \chi_R \cdot \chi_\Gamma$$

$$n_{A1'} = \frac{1}{12}[1\times1\times12 + 0 + 3\times1\times(-2) + 1\times1\times4 + 2\times1\times(-2) + 3\times1\times2] = 1$$

$$n_{A2'} = \frac{1}{12}[1\times1\times12 + 0 + 3\times(-1)\times(-2) + 1\times1\times4 + 2\times1\times(-2) + 3\times(-1)\times2] = 1$$

$$n_{E'} = \frac{1}{12}[1\times2\times12 + 0 + 0 + 1\times2\times4 + 2\times(-1)\times(-2) + 0] = 3$$

$$n_{A1''} = \frac{1}{12}[1\times1\times12 + 0 + 3\times1\times(-2) + (-4) + 2\times(-1)\times(-2) + (-6)] = 0$$

$$n_{A2''} = \frac{1}{12}[1\times1\times12 + 0 + 3\times(-1)\times(-2) + (-4) + 2\times(-1)\times(-2) + 6] = 2$$

$$n_{E''} = \frac{1}{12}[1\times1\times12 + 0 + 0 + 1\times(-2)\times4 + 2\times1\times(-2) + 0] = 1$$

$\rightarrow \Gamma_{Cartesian} = A_1' + A_2' + 3E' + 2A_2'' + E''$

$\Gamma_{Vibration} = A_1' + A_2' + 3E' + 2A_2'' + E'' - [E' + A_2'' + A_2' + E''] = A_1' + 2E' + A_2''$

Exercise 3: Determine the number and symmetry of vibrational modes in the N_2F_2 trans molecule.

$$F-N{=}N-F$$

Solution: This molecule belongs to the C_{2h} point group and since it is a non-linear four-atom molecule, its number of degrees of freedom will be 3N-6.

Symmetrical Elements	χ	The number of atoms not displaced	Γ_{xyz}
E	3	4	12
C_2	-1	0	0
i	-3	0	0
σ_h	1	4	4

Symbol Group C_{2h} Character Table:

C_{2H}	E	C_2	i	σ_h	Linear bases
A_g	1	1	1	1	R_z
B_G	1	-1	1	-1	R_x, R_y
A_u	1	1	-1	-1	Z
B_u	1	-1	-1	1	x,y
$\Gamma_{cartesian}$	12	0	0	4	

We reduce the $\Gamma_{cartesian}$:

$$n_\Gamma = \frac{1}{h} \Sigma_g \, n_g \cdot \chi_R \cdot \chi_\Gamma$$

$$n_{Ag} = \frac{1}{4}[1 \times 1 \times 12 + 0 + 0 + 1 \times 1 \times 4] = 4$$

$$n_{Bg} = \frac{1}{4}[1 \times 1 \times 12 + 0 + 0 + 1 \times (\text{-}1) \times 4] = 2$$

$$n_{Au} = \frac{1}{4}[1 \times 1 \times 12 + 0 + 0 + 1 \times (\text{-}1) \times 4] = 2$$

$$n_{Bu} = \frac{1}{4}[1 \times 1 \times 12 + 0 + 0 + 1 \times 1 \times 4] = 4$$

$$\rightarrow \Gamma_{Cartesian} = 4A_g + 2B_g + 2A_u + 4B_u$$

$$\rightarrow \Gamma_{Cartesian} = \Gamma_{Transitional} + \Gamma_{Rotate} + \Gamma_{Vibrating}$$

$$\Gamma_{Transitional} = A_u + 2B_u$$

$$\Gamma_{Rotate} = A_g + 2B_g$$

$$\Gamma_{Vibrating} = \Gamma_{Cartesian} - [\Gamma_{Transitional} + \Gamma_{Rotate}]$$

$$\rightarrow \Gamma_{Vibrating} = 4A_g + 2B_g + 2A_u + 4B_u - [A_u + 2B_u + A_g + 2B_g]$$

$$\rightarrow \Gamma_{Vibrating} = 3A_g + A_u + 2B_u$$

Exercise 4: Determine the number and symmetry of vibrational modes in the ammonia molecule.

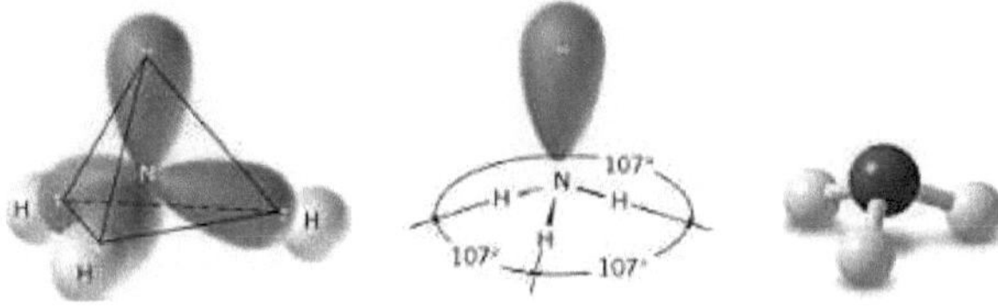

Solution: The ammonia molecule belongs to the C_{3v} point group and since it is a nonlinear molecule, its number of degrees of freedom will be 3N-6.

Symmetrical Elements	χ	The number of atoms not displaced	Γ_{xyz}
E	3	4	12
$2C_3$	0	1	0
$3\sigma_v$	1	2	2

Symbol Group C_{2h} Character Table:

C_{3v}	E	$2C_3$	$3\sigma_v$	Linear bases
A_1	1	1	1	Z
A_2	1	1	-1	R_z
E	2	-1	0	$(R_x, R_y), (x,y)$
$\Gamma_{Cartesian}$	12	0	2	
Γ_r	3	0	1	
Γ_α	3	0	1	

$$n_\Gamma = \frac{1}{h}\Sigma_g\, n_g \cdot \chi_R \cdot \chi_\Gamma$$

$$n_{A1} = \frac{1}{6}[1 \times 1 \times 12 + 0 + 3 \times 1 \times 2] = 3$$

$$n_{A2} = \frac{1}{6}[1 \times 1 \times 12 + 0 + 3 \times (-1) \times 2] = 1$$

$$n_E = \frac{1}{6}[1 \times 2 \times 12 + 0 + 0] = 4$$

$$\rightarrow \Gamma_{Cartesian} = 3A_1 + A_2 + 4E$$

$$\rightarrow \Gamma_{Cartesian} = \Gamma_{Transitional} + \Gamma_{Rotate} + \Gamma_{Vibartion}$$

$$\Gamma_{Transitional} = A_1 + E$$

$$\Gamma_{Rotate} = A_2 + E$$

$$\Gamma_{Vibartion} = \Gamma_{Cartesian} - [\Gamma_{Transitional} + \Gamma_{Rotate}]$$

$$\rightarrow \Gamma_{Vibartion} = 3A_1 + A_2 + 4E - [A_1 + A_2 + 2E]$$

$$\rightarrow \Gamma_{Vibartion} = 2A_1 + 2E$$

Now let's reduce the values of Γ_r and Γ_A:

$$n_{A1} = \frac{1}{6}[1 \times 1 \times 3 + 0 + 3 \times 1 \times 1] = 1$$

$$n_{A2} = \frac{1}{6}[1 \times 1 \times 3 + 0 + 3 \times (-1) \times 1] = 0$$

$$n_E = \frac{1}{6}[1 \times 2 \times 3 + 0 + 0] = 1$$

$$\Gamma_R = \Gamma_A = A_1 + E$$

Note 1: If the symbols x, y, z in the character table are written in the column opposite to the irreducible display, those displays are active in IR and if binary symbols x_y, x_z, y_z are written, those displays are in active Raman spectrum they will be.

Note 2: In molecules having the center of symmetry (i) vibrations that are active in Raman are not active in IR and vice versa.

Exercise 5: Find the vibrational modes in the CH_4 molecule.

Solution: This molecule belongs to the T_d point group and since it is a non-linear five-atom molecule, its number of degrees of freedom will be 3N-6 = 9.

SYMMETRICAL ELEMENTS	X	The number of atoms not displaced	Γ_{xyz}
E	3	5	15
C_3	0	2	0
C_2	-1	1	-1
S_4	-1	1	-1
σ_d	1	3	3

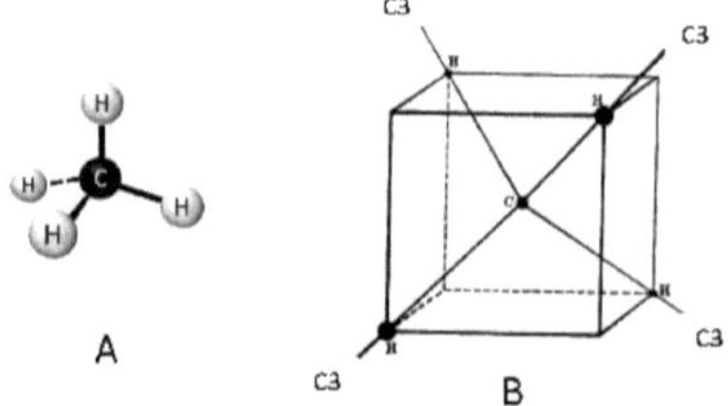

Figure19. Methane (A), C_3 axes in Methane (B)

T_d	E	$8C_3$	$3C_2$	$6S_4$	$6\sigma_d$	Linear bases
A_1	1	1	1	1	1	
A_2	1	1	1	-1	-1	
E	2	-1	2	0	0	
T_1	3	0	-1	1	-1	R_x, R_y, R_z
T_2	3	0	-1	-1	1	x, y, z
$\Gamma_{Cartesian}$	15	0	-1	-1	3	

Symbol Group Character Table T_d:

We reduce the $\Gamma_{Cartesian}$:

$$n\Gamma = \frac{1}{h} \Sigma_g \, n_g \cdot \chi_R \cdot \chi_\Gamma$$

$$n_{A1} = \frac{1}{24}[1 \times 1 \times 15 + 0 + 3 \times 1 \times (-1) + 6 \times 1 \times (-1) + 6 \times 1 \times 3] = 1$$

$$n_{A2} = \frac{1}{24}[1 \times 1 \times 15 + 0 + 3 \times 1 \times (-1) + 6 \times (-1) \times (-1) + 6 \times (-1) \times 3] = 0$$

$$n_E = \frac{1}{24}[1 \times 2 \times 15 + 0 + 3 \times 2 \times (-1) + 6 \times 0 \times (-1) + 6 \times 0 \times 3] = 1$$

$$n_{T1} = \frac{1}{24}[1 \times 3 \times 15 + 0 + 3 \times (-1) \times (-1) + 6 \times 1 \times (-1) + 6 \times (-1) \times 3] = 1$$

$$n_{T2} = \frac{1}{24}[1 \times 3 \times 15 + 0 + 3 \times (-1) \times (-1) + 6 \times (-1) \times (-1) + 6 \times 1 \times 3] = 3$$

$$\rightarrow \Gamma_{Cartesian} = A_1 + E + T_1 + 3T$$

$$\rightarrow \Gamma_{Cartesian} = \Gamma_{Transitional} + \Gamma_{Rotate} + \Gamma_{Vibrating}$$

$$\Gamma_{Transitional} = T_2$$

$$\Gamma_{Rotate} = T_1$$

$$\Gamma_{Vibrating} = \Gamma_{Cartesian} - [\Gamma_{Transitional} + \Gamma_{Rotate}]$$

$$\rightarrow \Gamma_{Vibrating} = A1 + E + T1 + 3T2 - [T2 + T1]$$

$$\rightarrow \Gamma_{Vibrating} = A1 + E + 2T2$$

Exercise 6: What vibrations are expected with symmetric components in the IR spectrum of the SO_2 gas molecule?

The SO_2 gas infrared spectrum has bands at 519, 606, 1151, 1361, 1871, 2304,

2499 cm^{-1}.

Solution: If we consider the molecule page yz:

SYMMETRICAL ELEMENTS	X	The number of atoms not displaced	Γ_{xyz}
E	3	3	9
$C_2(Z)$	-1	1	-1
σ_{xz}	1	1	1
σ_{yz}	1	3	3

After solving the C_{2v} point group character table, we obtain the vibrational value $\Gamma_{Vibrating}$:

$$\Gamma_{Vibrating} = 2A_1 + B_2$$

Since flexural vibrations have lower energy, then A_1 corresponds to $\upsilon_2 = 519$ cm^{-1}; and since asymmetric tensile vibrations have higher energy than symmetric tensile vibrations, then: $\upsilon_1 = 1151$ cm^{-1}, $\upsilon_3 = 1361$ cm^{-1}

Flexural Vibrations< Symmetric Tensile Vibrations< Asymmetric Tensile Vibrations

$\upsilon_2 + \upsilon_3 = 519 + 1361 = 1880 \approx 1871$ cm^{-1*}

$\upsilon_1 + \upsilon_3 = 1151 + 1361 = 2512 \approx 2499$ cm^{-1}

$\upsilon_1 - \upsilon_2 = 1151 - 519 = 632 \approx 606$ cm^{-1}

*These differences are due to the experimental error and the inconsistency of the vibrational modes.

In tensile vibrations sometimes a vibration can be combined with a vibration of its own or with other vibrations. The numbers obtained from this combination in theoretical calculations vary with experimental values of 20 to 30 units. Given that we consider the lowest energy vibration as bending vibration, the vibrations created by doubling a given vibration are called overtones.

The additional bands in the sub-red spectrum belong to the overtones and composite bands. For example, $2\upsilon_1$ is the first overtones υ_1.

$2\upsilon_1 = 2 \times 1151 = 2302 \approx 2304 \text{ cm}^{-1}$

If the vibrations match, the original vibrations can be obtained. For example:

$\upsilon_1 \rightarrow 2\upsilon_1 + 3\upsilon_1$

$$\upsilon_1 \pm \upsilon_2$$

$\upsilon_2 \rightarrow 2\upsilon_2 + 3\upsilon_2$ Combine two vibrations $\upsilon_2 \pm \upsilon_3$

$\upsilon_3 \rightarrow 2\upsilon_3 + 3\upsilon_3$ $\upsilon_1 \pm \upsilon_3$

Note: To determine whether or not Everton vibration is allowed or not, two main vibrations must be multiplied together. for example:

C_{2v}	E	$C_2(z)$	σ_{xz}	σ_{yz}
A_1	1	1	1	1
A_2	1	1	-1	-1
B_1	1	-1	1	-1
B_2	1	-1	-1	1
$B_2 = A_1 \times B_2$	1	-1	-1	1
$A_1 = A_1 \times A_2$	1	1	1	1
$A_1 = B_2 \times B_2$	1	1	1	1

$$A_1 + B_2 = \Gamma_{A1} \times \Gamma_{B2} = \Gamma_{B2}$$

$$A_1 + A_1 = \Gamma_{A1} \times \Gamma_{A1} = \Gamma_{A1}$$

$$B_2 + B_2 = \Gamma_{B2} \times \Gamma_{B2} = \Gamma_{A1}$$

Result: Since there are vibration effects in this set, vibration will be allowed.

 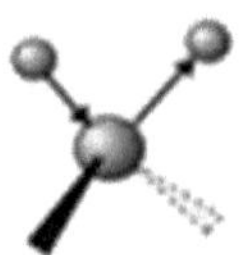

Character tables for important point groups

1. THE NONAXIAL GROUPS

C_1	E
A	1

C_s	E	σ_h		
A'	1	1	x, y, R_z	x^2, y^2, z^2, xy
A''	1	-1	z, R_x, R_y	yz, xz

C_i	E	i		
A_g	1	1	R_x, R_y, R_z	$x^2, y^2, z^2, xy, xz, yz$
A_u	1	-1	x, y, z	

2. THE C_n GROUPS

C_2	E	C_2		
A	1	1	z, R_z	x^2, y^2, z^2, xy
B	1	-1	x, y, R_x, R_y	yz, xz

C_3	E	C_3	C_3^2		$\varepsilon = \exp(2\pi i/3)$
A	1	1	1	z, R_z	$x^2 + y^2, z^2$
E	$\left\{\begin{matrix}1 \\ 1\end{matrix}\right.$ $\begin{matrix}\varepsilon \\ \varepsilon^*\end{matrix}$ $\left.\begin{matrix}\varepsilon^* \\ \varepsilon\end{matrix}\right\}$			$(x, y)(R_x, R_y)$	$(x^2 - y^2, xy)(yz, xz)$

C_4	E	C_4	C_2	C_4^3		
A	1	1	1	1	z, R_z	$x^2 + y^2, z^2$
B	1	-1	1	-1		$x^2 - y^2, xy$
E	$\left\{\begin{matrix}1 \\ 1\end{matrix}\right.$ $\begin{matrix}i \\ -i\end{matrix}$ $\begin{matrix}-1 \\ -1\end{matrix}$ $\left.\begin{matrix}-i \\ i\end{matrix}\right\}$				$(x, y)(R_x, R_y)$	(yz, xz)

2. THE C_n GROUPS (*continued*)

C_5	E	C_5	C_5^2	C_5^3	C_5^4		$\varepsilon = \exp(2\pi i/5)$
A	1	1	1	1	1	z, R_z	$x^2 + y^2, z^2$
E_1	1	ε	ε^2	ε^{2*}	ε^*	$(x, y)(R_x, R_y)$	(yz, xz)
	1	ε^*	ε^{2*}	ε^2	ε		
E_2	1	ε^2	ε^*	ε	ε^{2*}		$(x^2 - y^2, xy)$
	1	ε^{2*}	ε	ε^*	ε^2		

C_6	E	C_6	C_3	C_2	C_3^2	C_6^5		$\varepsilon = \exp(2\pi i/6)$
A	1	1	1	1	1	1	z, R_z	$x^2 + y^2, z^2$
B	1	-1	1	-1	1	-1		
E_1	1	ε	$-\varepsilon^*$	-1	$-\varepsilon$	ε^*	(x, y) (R_x, R_y)	(xz, yz)
	1	ε^*	$-\varepsilon$	-1	$-\varepsilon^*$	ε		
E_2	1	$-\varepsilon^*$	$-\varepsilon$	1	$-\varepsilon^*$	$-\varepsilon$		$(x^2 - y^2, xy)$
	1	$-\varepsilon$	$-\varepsilon^*$	1	$-\varepsilon$	$-\varepsilon^*$		

C_7	E	C_7	C_7^2	C_7^3	C_7^4	C_7^5	C_7^6		$\varepsilon = \exp(2\pi i/7)$
A	1	1	1	1	1	1	1	z, R_z	$x^2 + y^2, z^2$
E_1	1	ε	ε^2	ε^3	ε^{3*}	ε^{2*}	ε^*	(x, y) (R_x, R_y)	(xz, yz)
	1	ε^*	ε^{2*}	ε^{3*}	ε^3	ε^2	ε		
E_2	1	ε^2	ε^{3*}	ε^*	ε	ε^3	ε^{2*}		$(x^2 - y^2, xy)$
	1	ε^{2*}	ε^3	ε	ε^*	ε^{3*}	ε^2		
E_3	1	ε^3	ε^*	ε^2	ε^{2*}	ε	ε^{3*}		
	1	ε^{3*}	ε	ε^{2*}	ε^2	ε^*	ε^3		

C_8	E	C_8	C_4	C_2	C_4^3	C_8^3	C_8^5	C_8^7		$\varepsilon = \exp(2\pi i/8)$
A	1	1	1	1	1	1	1	1	z, R_z	$x^2 + y^2, z^2$
B	1	-1	1	1	1	-1	-1	-1		
E_1	1	ε	i	-1	$-i$	$-\varepsilon^*$	$-\varepsilon$	ε^*	(x, y) (R_x, R_y)	(xz, yz)
	1	ε^*	$-i$	-1	i	$-\varepsilon$	$-\varepsilon^*$	ε		
E_2	1	i	-1	1	-1	$-i$	i	$-i$		$(x^2 - y^2, xy)$
	1	$-i$	-1	1	-1	i	$-i$	i		
E_3	1	$-\varepsilon^*$	i	-1	$-i$	ε^*	ε	$-\varepsilon^*$		
	1	$-\varepsilon^*$	$-i$	-1	i	ε	ε^*	$-\varepsilon$		

3. THE D_n GROUPS

D_2	E	$C_3(z)$	$C_2(y)$	$C_2(x)$		
A	1	1	1	1		$x^2 + y^2, z^2$
B_1	1	1	-1	-1	z, R_z	xy
B_2	1	-1	1	-1	y, R_y	xz
B_3	1	-1	-1	1	x, R_x	yz

D_3	E	$2C_3$	$3C_2$		
A_1	1	1	1		$x^2 + y^2, z^2$
A_2	1	1	-1	z, R_z	
E	2	-1	0	$(x, y)(R_x, R_y)$	$(x^2 - y^2, xy)(xz, yz)$

D_4	E	$2C_4$	$C_2(=C_4{}^2)$	$2C_2'$	$2C_2''$		
A_1	1	1	1	1	1		$x^2 + y^2, z^2$
A_2	1	1	1	-1	-1	z, R_z	
B_1	1	-1	1	1	-1		$x^2 - y^2$
B_2	1	-1	1	-1	1		xy
E	2	0	-2	0	0	$(x, y)(R_x, R_y)$	(xz, yz)

3. THE D_n GROUPS (continued)

D_5	E	$2C_5$	$2C_5{}^2$	$5C_2$		
A_1	1	1	1	1		$x^2 + y^2, z^2$
A_2	1	1	1	-1	z, R_z	
E_1	2	$2 \cos 72°$	$2 \cos 144°$	0	$(x, y)(R_x, R_y)$	(xz, yz)
E_2	2	$2 \cos 144°$	$2 \cos 72°$	0		$(x^2 - y^2, xy)$

D_6	E	$2C_6$	$2C_3$	C_2	$3C_2'$	$3C_2''$		
A_1	1	1	1	1	1	1		$x^2 + y^2, z^2$
A_2	1	1	1	1	-1	-1	z, R_z	
B_1	1	-1	1	-1	1	-1		
B_2	1	-1	1	-1	-1	1		
E_1	2	1	-1	-2	0	0	$(x, y)(R_x, R_y)$	(xz, yz)
E_2	2	-1	-1	2	0	0		$(x^2 - y^2, xy)$

4. THE C_{nv} GROUPS

$C_{2v}{}^a$	E	C_2	$\sigma_v(xz)$	$\sigma_v'(yz)$		
A_1	1	1	1	1	z	x^2, y^2, z^2
A_2	1	1	-1	-1	R_z	xy
B_1	1	-1	1	-1	x, R_y	xz
B_2	1	-1	-1	1	y, R_x	yz

^{a} For a planar molecule the x-axis is taken perpendicular to the plane.

C_{3v}	E	$2C_3$	$3\sigma_v$		
A_1	1	1	1	z	$x^2 + y^2, z^2$
A_2	1	1	-1	R_z	
E	2	-1	0	$(x, y)(R_x, R_y)$	$(x^2 - y^2, xy)(xz, yz)$

$C_{4v}{}^a$	E	$2C_4$	C_2	$2\sigma_v$	$2\sigma_d$		
A_1	1	1	1	1	1	z	$x^2 + y^2, z^2$
A_2	1	1	1	-1	-1	R_z	
B_1	1	-1	1	1	-1		$x^2 - y^2$
B_2	1	-1	1	-1	1		xy
E	2	0	-2	0	0	$(x, y)(R_x, R_y)$	(xz, yz)

C_{5v}	E	$2C_5$	$2C_5^2$	$5\sigma_v$		
A_1	1	1	1	1	z	$x^2 + y^2,\ z^2$
A_2	1	1	1	-1	R_z	
E_1	2	$2\cos 72°$	$2\cos 144°$	0	$(x, y)(R_x, R_y)$	(xz, yz)
E_2	2	$2\cos 144°$	$2\cos 72°$	0		$(x^2 - y^2,\ xy)$

C_{6v}	E	$2C_6$	$2C_3$	C_2	$3\sigma_v$	$3\sigma_d$		
A_1	1	1	1	1	1	1	z	$x^2 + y^2,\ z^2$
A_2	1	1	1	1	-1	-1	R_z	
B_1	1	-1	1	-1	1	-1		
B_2	1	-1	1	-1	-1	1		
E_1	2	1	-1	-2	0	0	$(x, y)(R_x, R_y)$	(xz, yz)
E_2	2	-1	-1	2	0	0		$(x^2 - y^2,\ xy)$

5. THE C_{nh} GROUPS

C_{2h}	E	C_2	i	σ_h		
A_g	1	1	1	1	R_z	$x^2,\ y^2,\ z^2,\ xy$
B_g	1	-1	1	-1	$R_x,\ R_y$	$xz,\ yz$
A_u	1	1	-1	-1	z	
B_u	1	-1	-1	1	$x,\ y$	

C_{3h}	E	C_3	C_3^2	σ_h	S_3	S_3^5		$\varepsilon = \exp(2\pi i/3)$
A'	1	1	1	1	1	1	R_z	$x^2 + y^2,\ z^2$
E'	$\begin{cases}1\\1\end{cases}$	$\begin{matrix}\varepsilon\\\varepsilon^*\end{matrix}$	$\begin{matrix}\varepsilon^*\\\varepsilon\end{matrix}$	$\begin{matrix}1\\1\end{matrix}$	$\begin{matrix}\varepsilon\\\varepsilon^*\end{matrix}$	$\begin{matrix}\varepsilon^*\\\varepsilon\end{matrix}$	(x, y)	$(x^2 - y^2,\ xy)$
A''	1	1	1	-1	-1	-1	z	
E''	$\begin{cases}1\\1\end{cases}$	$\begin{matrix}\varepsilon\\\varepsilon^*\end{matrix}$	$\begin{matrix}\varepsilon^*\\\varepsilon\end{matrix}$	$\begin{matrix}-1\\-1\end{matrix}$	$\begin{matrix}-\varepsilon\\-\varepsilon^*\end{matrix}$	$\begin{matrix}-\varepsilon^*\\-\varepsilon\end{matrix}$	(R_x, R_y)	(xz, yz)

C_{4h}	E	C_4	C_2	C_4^3	i	S_4^3	σ_h	S_4		
A_g	1	1	1	1	1	1	1	1	R_z	$x^2 + y^2,\ z^2$
B_g	1	-1	1	-1	1	-1	1	-1		$x^2 - y^2,\ xy$
E_g	$\begin{cases}1\\1\end{cases}$	$\begin{matrix}i\\-i\end{matrix}$	$\begin{matrix}-1\\-1\end{matrix}$	$\begin{matrix}-i\\i\end{matrix}$	$\begin{matrix}1\\1\end{matrix}$	$\begin{matrix}i\\-i\end{matrix}$	$\begin{matrix}-1\\-1\end{matrix}$	$\begin{matrix}-i\\i\end{matrix}$	(R_x, R_y)	(xz, yz)
A_u	1	1	1	1	-1	-1	-1	-1	z	
B_u	1	-1	1	-1	-1	1	-1	1		
E_u	$\begin{cases}1\\1\end{cases}$	$\begin{matrix}i\\-i\end{matrix}$	$\begin{matrix}-1\\-1\end{matrix}$	$\begin{matrix}-i\\i\end{matrix}$	$\begin{matrix}-1\\-1\end{matrix}$	$\begin{matrix}-i\\i\end{matrix}$	$\begin{matrix}1\\1\end{matrix}$	$\begin{matrix}i\\-i\end{matrix}$	(x, y)	

C_{5h}	E	C_5	C_5^2	C_5^3	C_5^4	σ_h	S_5	S_5^7	S_5^3	S_5^9		$\varepsilon = \exp(2\pi i/5)$
A'	1	1	1	1	1	1	1	1	1	1	R_z	$x^2+y^2,\ z^2$
E_1'	$\left\{\begin{matrix}1\\1\end{matrix}\right.$	$\begin{matrix}\varepsilon\\\varepsilon^*\end{matrix}$	$\begin{matrix}\varepsilon^2\\\varepsilon^{2*}\end{matrix}$	$\begin{matrix}\varepsilon^{2*}\\\varepsilon^2\end{matrix}$	$\begin{matrix}\varepsilon^*\\\varepsilon\end{matrix}$	$\begin{matrix}1\\1\end{matrix}$	$\begin{matrix}\varepsilon\\\varepsilon^*\end{matrix}$	$\begin{matrix}\varepsilon^2\\\varepsilon^{2*}\end{matrix}$	$\begin{matrix}\varepsilon^{2*}\\\varepsilon^2\end{matrix}$	$\left.\begin{matrix}\varepsilon^*\\\varepsilon\end{matrix}\right\}$	(x, y)	
E_2'	$\left\{\begin{matrix}1\\1\end{matrix}\right.$	$\begin{matrix}\varepsilon^2\\\varepsilon^{2*}\end{matrix}$	$\begin{matrix}\varepsilon^*\\\varepsilon\end{matrix}$	$\begin{matrix}\varepsilon\\\varepsilon^*\end{matrix}$	$\begin{matrix}\varepsilon^{2*}\\\varepsilon^2\end{matrix}$	$\begin{matrix}1\\1\end{matrix}$	$\begin{matrix}\varepsilon^2\\\varepsilon^{2*}\end{matrix}$	$\begin{matrix}\varepsilon^*\\\varepsilon\end{matrix}$	$\begin{matrix}\varepsilon\\\varepsilon^*\end{matrix}$	$\left.\begin{matrix}\varepsilon^{2*}\\\varepsilon^2\end{matrix}\right\}$		$(x^2-y^2,\ xy)$
A''	1	1	1	1	1	-1	-1	-1	-1	-1	z	
E_1''	$\left\{\begin{matrix}1\\1\end{matrix}\right.$	$\begin{matrix}\varepsilon\\\varepsilon^*\end{matrix}$	$\begin{matrix}\varepsilon^2\\\varepsilon^{2*}\end{matrix}$	$\begin{matrix}\varepsilon^{2*}\\\varepsilon^2\end{matrix}$	$\begin{matrix}\varepsilon^*\\\varepsilon\end{matrix}$	$\begin{matrix}-1\\-1\end{matrix}$	$\begin{matrix}-\varepsilon\\-\varepsilon^*\end{matrix}$	$\begin{matrix}-\varepsilon^2\\-\varepsilon^{2*}\end{matrix}$	$\begin{matrix}-\varepsilon^{2*}\\-\varepsilon^2\end{matrix}$	$\left.\begin{matrix}-\varepsilon^*\\-\varepsilon\end{matrix}\right\}$	(R_x, R_y)	(xz, yz)
E_2''	$\left\{\begin{matrix}1\\1\end{matrix}\right.$	$\begin{matrix}\varepsilon^2\\\varepsilon^{2*}\end{matrix}$	$\begin{matrix}\varepsilon^*\\\varepsilon\end{matrix}$	$\begin{matrix}\varepsilon\\\varepsilon^*\end{matrix}$	$\begin{matrix}\varepsilon^{2*}\\\varepsilon^2\end{matrix}$	$\begin{matrix}-1\\-1\end{matrix}$	$\begin{matrix}-\varepsilon^2\\-\varepsilon^{2*}\end{matrix}$	$\begin{matrix}-\varepsilon^*\\-\varepsilon\end{matrix}$	$\begin{matrix}-\varepsilon\\-\varepsilon^*\end{matrix}$	$\left.\begin{matrix}-\varepsilon^{2*}\\-\varepsilon^2\end{matrix}\right\}$		

C_{6h}	E	C_6	C_3	C_2	C_3^2	C_6^5	i	S_3^5	S_6^5	σ_h	S_6	S_3		$\varepsilon = \exp(2\pi i/6)$
A_g	1	1	1	1	1	1	1	1	1	1	1	1	R_z	$x^2+y^2,\ z^2$
B_g	1	-1	1	-1	1	-1	1	-1	1	-1	1	-1		
E_{1g}	$\left\{\begin{matrix}1\\1\end{matrix}\right.$	$\begin{matrix}\varepsilon\\\varepsilon^*\end{matrix}$	$\begin{matrix}-\varepsilon^*\\-\varepsilon\end{matrix}$	$\begin{matrix}-1\\-1\end{matrix}$	$\begin{matrix}-\varepsilon\\-\varepsilon^*\end{matrix}$	$\begin{matrix}\varepsilon^*\\\varepsilon\end{matrix}$	$\begin{matrix}1\\1\end{matrix}$	$\begin{matrix}\varepsilon\\\varepsilon^*\end{matrix}$	$\begin{matrix}-\varepsilon^*\\-\varepsilon\end{matrix}$	$\begin{matrix}-1\\-1\end{matrix}$	$\begin{matrix}-\varepsilon\\-\varepsilon^*\end{matrix}$	$\left.\begin{matrix}\varepsilon^*\\\varepsilon\end{matrix}\right\}$	(R_x, R_y)	(xz, yz)
E_{2g}	$\left\{\begin{matrix}1\\1\end{matrix}\right.$	$\begin{matrix}-\varepsilon^*\\-\varepsilon\end{matrix}$	$\begin{matrix}-\varepsilon\\-\varepsilon^*\end{matrix}$	$\begin{matrix}1\\1\end{matrix}$	$\begin{matrix}-\varepsilon^*\\-\varepsilon\end{matrix}$	$\begin{matrix}-\varepsilon\\-\varepsilon^*\end{matrix}$	$\begin{matrix}1\\1\end{matrix}$	$\begin{matrix}-\varepsilon^*\\-\varepsilon\end{matrix}$	$\begin{matrix}-\varepsilon\\-\varepsilon^*\end{matrix}$	$\begin{matrix}1\\1\end{matrix}$	$\begin{matrix}-\varepsilon^*\\-\varepsilon\end{matrix}$	$\left.\begin{matrix}-\varepsilon\\-\varepsilon^*\end{matrix}\right\}$		$(x^2-y^2,\ xy)$
A_u	1	1	1	1	1	1	-1	-1	-1	-1	-1	-1	z	
B_u	1	-1	1	-1	1	-1	-1	1	-1	1	-1	1		
E_{1u}	$\left\{\begin{matrix}1\\1\end{matrix}\right.$	$\begin{matrix}\varepsilon\\\varepsilon^*\end{matrix}$	$\begin{matrix}-\varepsilon^*\\-\varepsilon\end{matrix}$	$\begin{matrix}-1\\-1\end{matrix}$	$\begin{matrix}-\varepsilon\\-\varepsilon^*\end{matrix}$	$\begin{matrix}\varepsilon^*\\\varepsilon\end{matrix}$	$\begin{matrix}-1\\-1\end{matrix}$	$\begin{matrix}-\varepsilon\\-\varepsilon^*\end{matrix}$	$\begin{matrix}\varepsilon^*\\\varepsilon\end{matrix}$	$\begin{matrix}1\\1\end{matrix}$	$\begin{matrix}\varepsilon\\\varepsilon^*\end{matrix}$	$\left.\begin{matrix}-\varepsilon^*\\-\varepsilon\end{matrix}\right\}$	(x, y)	
E_{2u}	$\left\{\begin{matrix}1\\1\end{matrix}\right.$	$\begin{matrix}-\varepsilon^*\\-\varepsilon\end{matrix}$	$\begin{matrix}-\varepsilon\\-\varepsilon^*\end{matrix}$	$\begin{matrix}1\\1\end{matrix}$	$\begin{matrix}-\varepsilon^*\\-\varepsilon\end{matrix}$	$\begin{matrix}-\varepsilon\\-\varepsilon^*\end{matrix}$	$\begin{matrix}-1\\-1\end{matrix}$	$\begin{matrix}\varepsilon^*\\\varepsilon\end{matrix}$	$\begin{matrix}\varepsilon\\\varepsilon^*\end{matrix}$	$\begin{matrix}-1\\-1\end{matrix}$	$\begin{matrix}\varepsilon^*\\\varepsilon\end{matrix}$	$\left.\begin{matrix}\varepsilon\\\varepsilon^*\end{matrix}\right\}$		

6. THE D_{nh} GROUPS

D_{2h}	E	$C_2(z)$	$C_2(y)$	$C_2(x)$	i	$\sigma(xy)$	$\sigma(xz)$	$\sigma(yz)$		
A_g	1	1	1	1	1	1	1	1		$x^2,\ y^2,\ z^2$
B_{1g}	1	1	-1	-1	1	1	-1	-1	R_z	xy
B_{2g}	1	-1	1	-1	1	-1	1	-1	R_y	xz
B_{3g}	1	-1	-1	1	1	-1	-1	1	R_x	yz
A_u	1	1	1	1	-1	-1	-1	-1		
B_{1u}	1	1	-1	-1	-1	-1	1	1	z	
B_{2u}	1	-1	1	-1	-1	1	-1	1	y	
B_{3u}	1	-1	-1	1	-1	1	1	-1	x	

D_{3h}	E	$2C_3$	$3C_2$	σ_h	$2S_3$	$3\sigma_v$		
A_1'	1	1	1	1	1	1		$x^2+y^2,\ z^2$
A_2'	1	1	-1	1	1	-1	R_z	
E'	2	-1	0	2	-1	0	(x, y)	$(x^2-y^2,\ xy)$
A_1''	1	1	1	-1	-1	-1		
A_2''	1	1	-1	-1	-1	1	z	
E''	2	-1	0	-2	1	0	(R_x, R_y)	(xz, yz)

$D_{4h}{}^a$	E	$2C_4$	C_2	$2C_2'$	$2C_2''$	i	$2S_4$	σ_h	$2\sigma_v$	$2\sigma_d$		
A_{1g}	1	1	1	1	1	1	1	1	1	1		$x^2 + y^2, z^2$
A_{2g}	1	1	1	-1	-1	1	1	1	-1	-1	R_z	
B_{1g}	1	-1	1	1	-1	1	-1	1	1	-1		$x^2 - y^2$
B_{2g}	1	-1	1	-1	1	1	-1	1	-1	1		xy
E_g	2	0	-2	0	0	2	0	-2	0	0	(R_x, R_y)	(xz, yz)
A_{1u}	1	1	1	1	1	-1	-1	-1	-1	-1		
A_{2u}	1	1	1	-1	-1	-1	-1	-1	1	1	z	
B_{1u}	1	-1	1	1	-1	-1	1	-1	-1	1		
B_{2u}	1	-1	1	-1	1	-1	1	-1	1	-1		
E_u	2	0	-2	0	0	-2	0	2	0	0	(x, y)	

$^a\sigma_v$ passes through the atoms and σ_d bisects the bond angles.

D_{5h}	E	$2C_5$	$2C_5{}^2$	$5C_2$	σ_h	$2S_5$	$2S_5{}^3$	$5\sigma_v$		
A_1'	1	1	1	1	1	1	1	1		$x^2 + y^2, z^2$
A_2'	1	1	1	-1	1	1	1	-1	R_z	
E_1'	2	$2\cos 72°$	$2\cos 144°$	0	2	$2\cos 72°$	$2\cos 144°$	0	(x, y)	
E_2'	2	$2\cos 144°$	$2\cos 72°$	0	2	$2\cos 144°$	$2\cos 72°$	0		$(x^2 - y^2, xy)$
A_1''	1	1	1	1	-1	-1	-1	-1		
A_2''	1	1	1	-1	-1	-1	-1	1	z	
E_1''	2	$2\cos 72°$	$2\cos 144°$	0	-2	$-2\cos 72°$	$-2\cos 144°$	0	(R_x, R_y)	(xz, yz)
E_2''	2	$2\cos 144°$	$2\cos 72°$	0	-2	$-2\cos 144°$	$-2\cos 72°$	0		

C_{5h}	E	C_5	C_5^2	C_5^3	C_5^4	σ_h	S_5	S_5^7	S_5^3	S_5^9		$\varepsilon = \exp(2\pi i/5)$
A'	1	1	1	1	1	1	1	1	1	1	R_z	$x^2+y^2,\ z^2$
E_1'	$\{1$	ε	ε^2	ε^{2*}	ε^*	1	ε	ε^2	ε^{2*}	$\varepsilon^*\}$	(x, y)	
	$\{1$	ε^*	ε^{2*}	ε^2	ε	1	ε^*	ε^{2*}	ε^2	$\varepsilon\}$		
E_2'	$\{1$	ε^2	ε^*	ε	ε^{2*}	1	ε^2	ε^*	ε	$\varepsilon^{2*}\}$		$(x^2-y^2,\ xy)$
	$\{1$	ε^{2*}	ε	ε^*	ε^2	1	ε^{2*}	ε	ε^*	$\varepsilon^2\}$		
A''	1	1	1	1	1	-1	-1	-1	-1	-1	z	
E_1''	$\{1$	ε	ε^2	ε^{2*}	ε^*	-1	$-\varepsilon$	$-\varepsilon^2$	$-\varepsilon^{2*}$	$-\varepsilon^*\}$	(R_x, R_y)	(xz, yz)
	$\{1$	ε^*	ε^{2*}	ε^2	ε	-1	$-\varepsilon^*$	$-\varepsilon^{2*}$	$-\varepsilon^2$	$-\varepsilon\}$		
E_2''	$\{1$	ε^2	ε^*	ε	ε^{2*}	-1	$-\varepsilon^2$	$-\varepsilon^*$	$-\varepsilon$	$-\varepsilon^{2*}\}$		
	$\{1$	ε^{2*}	ε	ε^*	ε^2	-1	$-\varepsilon^{2*}$	$-\varepsilon$	$-\varepsilon^*$	$-\varepsilon^2\}$		

C_{6h}	E	C_6	C_3	C_2	C_3'	C_6'	i	S_3'	S_6'	σ_h	S_6	S_3		$\varepsilon = \exp(2\pi i/6)$
A_g	1	1	1	1	1	1	1	1	1	1	1	1	R_z	$x^2+y^2,\ z^2$
B_g	1	-1	1	-1	1	-1	1	-1	1	-1	1	-1		
E_{1g}	$\{1$	ε	$-\varepsilon^*$	-1	$-\varepsilon$	ε^*	1	ε	$-\varepsilon^*$	-1	$-\varepsilon$	$\varepsilon^*\}$	(R_x, R_y)	(xz, yz)
	$\{1$	ε^*	$-\varepsilon$	-1	$-\varepsilon^*$	ε	1	ε^*	$-\varepsilon$	-1	$-\varepsilon^*$	$\varepsilon\}$		
E_{2g}	$\{1$	$-\varepsilon^*$	$-\varepsilon$	1	$-\varepsilon^*$	$-\varepsilon$	1	$-\varepsilon^*$	$-\varepsilon$	1	$-\varepsilon^*$	$-\varepsilon\}$		$(x^2-y^2,\ xy)$
	$\{1$	$-\varepsilon$	$-\varepsilon^*$	1	$-\varepsilon$	$-\varepsilon^*$	1	$-\varepsilon$	$-\varepsilon^*$	1	$-\varepsilon$	$-\varepsilon^*\}$		
A_u	1	1	1	1	1	1	-1	-1	-1	-1	-1	-1	z	
B_u	1	-1	1	-1	1	-1	-1	1	-1	1	-1	1		
E_{1u}	$\{1$	ε	$-\varepsilon^*$	-1	$-\varepsilon$	ε^*	-1	$-\varepsilon$	ε^*	1	ε	$-\varepsilon^*\}$	(x, y)	
	$\{1$	ε^*	$-\varepsilon$	-1	$-\varepsilon^*$	ε	-1	$-\varepsilon^*$	ε	1	ε^*	$-\varepsilon\}$		
E_{2u}	$\{1$	$-\varepsilon^*$	$-\varepsilon$	1	$-\varepsilon^*$	$-\varepsilon$	-1	ε^*	ε	-1	ε^*	$\varepsilon\}$		
	$\{1$	$-\varepsilon$	$-\varepsilon^*$	1	$-\varepsilon$	$-\varepsilon^*$	-1	ε	ε^*	-1	ε	$\varepsilon^*\}$		

6. THE D_{nh} GROUPS

D_{2h}	E	$C_2(z)$	$C_2(y)$	$C_2(x)$	i	$\sigma(xy)$	$\sigma(xz)$	$\sigma(yz)$		
A_g	1	1	1	1	1	1	1	1		x^2, y^2, z^2
B_{1g}	1	1	-1	-1	1	1	-1	-1	R_z	xy
B_{2g}	1	-1	1	-1	1	-1	1	-1	R_y	xz
B_{3g}	1	-1	-1	1	1	-1	-1	1	R_x	yz
A_u	1	1	1	1	-1	-1	-1	-1		
B_{1u}	1	1	-1	-1	-1	-1	1	1	z	
B_{2u}	1	-1	1	-1	-1	1	-1	1	y	
B_{3u}	1	-1	-1	1	-1	1	1	-1	x	

D_{3h}	E	$2C_3$	$3C_2$	σ_h	$2S_3$	$3\sigma_v$		
A_1'	1	1	1	1	1	1		$x^2+y^2,\ z^2$
A_2'	1	1	-1	1	1	-1	R_z	
E'	2	-1	0	2	-1	0	(x, y)	$(x^2-y^2,\ xy)$
A_1''	1	1	1	-1	-1	-1		
A_2''	1	1	-1	-1	-1	1	z	
E''	2	-1	0	-2	1	0	(R_x, R_y)	(xz, yz)

D_{6h}	E	$2C_6$	$2C_3$	C_2	$3C_2'$	$3C_2''$	i	$2S_3$	$2S_6$	σ_h	$3\sigma_d$	$3\sigma_v$		
A_{1g}	1	1	1	1	1	1	1	1	1	1	1	1		x^2+y^2, z^2
A_{2g}	1	1	1	1	-1	-1	1	1	1	1	-1	-1	R_z	
B_{1g}	1	-1	1	-1	1	-1	1	-1	1	-1	1	-1		
B_{2g}	1	-1	1	-1	-1	1	1	-1	1	-1	-1	1		
E_{1g}	2	1	-1	-2	0	0	2	1	-1	-2	0	0	(R_x, R_y)	(xz, yz)
E_{2g}	2	-1	-1	2	0	0	2	-1	-1	2	0	0		(x^2-y^2, xy)
A_{1u}	1	1	1	1	1	1	-1	-1	-1	-1	-1	-1		
A_{2u}	1	1	1	1	-1	-1	-1	-1	-1	-1	1	1	z	
B_{1u}	1	-1	1	-1	1	-1	-1	1	-1	1	-1	1		
B_{2u}	1	-1	1	-1	-1	1	-1	1	-1	1	1	-1		
E_{1u}	2	1	-1	-2	0	0	-2	-1	1	2	0	0	(x, y)	
E_{2u}	2	-1	-1	2	0	0	-2	1	1	-2	0	0		

6. THE D_{nh} GROUPS (continued)

D_{8h}	E	$2C_8$	$2C_8^3$	$2C_4$	C_2	$4C_2'$	$4C_2''$	i	$2S_8$	$2S_8^3$	$2S_4$	σ_h	$4\sigma_d$	$4\sigma_v$		
A_{1g}	1	1	1	1	1	1	1	1	1	1	1	1	1	1		x^2+y^2, z^2
A_{2g}	1	1	1	1	1	-1	-1	1	1	1	1	1	-1	-1	R_z	
B_{1g}	1	-1	-1	1	1	1	-1	1	-1	-1	1	1	1	-1		
B_{2g}	1	-1	-1	1	1	-1	1	1	-1	-1	1	1	-1	1		
E_{1g}	2	$\sqrt{2}$	$-\sqrt{2}$	0	-2	0	0	2	$\sqrt{2}$	$-\sqrt{2}$	0	-2	0	0	(R_x, R_y)	(xz, yz)
E_{2g}	2	0	0	-2	2	0	0	2	0	0	-2	2	0	0		(x^2-y^2, xy)
E_{3g}	2	$-\sqrt{2}$	$\sqrt{2}$	0	-2	0	0	2	$-\sqrt{2}$	$\sqrt{2}$	0	-2	0	0		
A_{1u}	1	1	1	1	1	1	1	-1	-1	-1	-1	-1	-1	-1		
A_{2u}	1	1	1	1	-1	-1	-1	-1	-1	-1	-1	-1	1	1	z	
B_{1u}	1	-1	-1	1	1	1	-1	-1	1	1	-1	-1	-1	1		
B_{2u}	1	-1	-1	1	1	-1	1	-1	1	1	-1	-1	1	-1		
E_{1u}	2	$\sqrt{2}$	$-\sqrt{2}$	0	-2	0	0	-2	$-\sqrt{2}$	$\sqrt{2}$	0	2	0	0	(x, y)	
E_{2u}	2	0	0	-2	2	0	0	-2	0	0	2	-2	0	0		
E_{3u}	2	$-\sqrt{2}$	$\sqrt{2}$	0	-2	0	0	-2	$\sqrt{2}$	$-\sqrt{2}$	0	2	0	0		

7. THE D_{nd} GROUPS

D_{2d}	E	$2S_4$	C_2	$2C_2'$	$2\sigma_d$		
A_1	1	1	1	1	1		$x^2+y^2,\ z^2$
A_2	1	1	1	-1	-1	R_z	
B_1	1	-1	1	1	-1		x^2-y^2
B_2	1	-1	1	-1	1	z	xy
E	1	0	-2	0	0	$(x,y),\ (R_x,R_y)$	(xz,yz)

D_{3d}	E	$2C_3$	$3C_2$	i	$2S_6$	$3\sigma_d$		
A_{1g}	1	1	1	1	1	1		$x^2+y^2,\ z^2$
A_{2g}	1	1	-1	1	1	-1	R_z	
E_g	2	-1	0	2	-1	0	(R_x,R_y)	(x^2-y^2,xy) (xz,yz)
A_{1u}	1	1	1	-1	-1	-1		
A_{2u}	1	1	-1	-1	-1	1	z	
E_u	2	-1	0	-2	1	0	(x,y)	

D_{4d}	E	$2S_8$	$2C_4$	$2S_8^3$	C_2	$4C_2'$	$4\sigma_d$		
A_1	1	1	1	1	1	1	1		$x^2+y^2,\ z^2$
A_2	1	1	1	1	1	-1	-1	R_z	
B_1	1	-1	1	-1	1	1	-1		
B_2	1	-1	1	-1	1	-1	1	z	
E_1	2	$\sqrt{2}$	0	$-\sqrt{2}$	-2	0	0	(x,y)	
E_2	2	0	-2	0	2	0	0		(x^2-y^2,xy)
E_3	2	$-\sqrt{2}$	0	$\sqrt{2}$	-2	0	0	(R_x,R_y)	(xz,yz)

D_{5d}	E	$2C_5$	$2C_5^2$	$5C_2$	i	$2S_{10}^3$	$2S_{10}$	$5\sigma_d$		
A_{1g}	1	1	1	1	1	1	1	1		$x^2+y^2,\ z^2$
A_{2g}	1	1	1	-1	1	1	1	-1	R_z	
E_{1g}	2	$2\cos 72°$	$2\cos 144°$	0	2	$2\cos 72°$	$2\cos 144°$	0	(R_x,R_y)	(xz,yz)
E_{2g}	2	$2\cos 144°$	$2\cos 72°$	0	2	$2\cos 144°$	$2\cos 72°$	0		(x^2-y^2,xy)
A_{1u}	1	1	1	1	-1	-1	-1	-1		
A_{2u}	1	1	1	-1	-1	-1	-1	1	z	
E_{1u}	2	$2\cos 72°$	$2\cos 144°$	0	-2	$-2\cos 72°$	$-2\cos 144°$	0	(x,y)	
E_{2u}	2	$2\cos 144°$	$2\cos 72°$	0	-2	$-2\cos 144°$	$-2\cos 72°$	0		

7. THE D_{nd} GROUPS (*continued*)

D_{6d}	E	$2S_{12}$	$2C_6$	$2S_4$	$2C_3$	$2S_{12}^{5}$	C_2	$6C_2'$	$6\sigma_d$		
A_1	1	1	1	1	1	1	1	1	1		$x^2+y^2,\ z^2$
A_2	1	1	1	1	1	1	1	-1	-1	R_z	
B_1	1	-1	1	-1	1	-1	1	1	-1		
B_2	1	-1	1	-1	1	-1	1	-1	1	z	
E_1	2	$\sqrt{3}$	1	0	-1	$-\sqrt{3}$	-2	0	0	(x, y)	
E_2	2	1	-1	-2	-1	1	2	0	0		$(x^2-y^2,\ xy)$
E_3	2	0	-2	0	2	0	-2	0	0		
E_4	2	-1	-1	2	-1	-1	2	0	0		
E_5	2	$-\sqrt{3}$	1	0	-1	$\sqrt{3}$	-2	0	0	(R_x, R_y)	(xz, yz)

8. THE S_n GROUPS

S_4	E	S_4	C_2	S_4^{3}		
A	1	1	1	1	R_z	$x^2+y^2,\ z^2$
B	1	-1	1	-1	z	$x^2-y^2,\ xy$
E	$\begin{Bmatrix}1\\1\end{Bmatrix}$	$\begin{matrix}i\\-i\end{matrix}$	$\begin{matrix}-1\\-1\end{matrix}$	$\begin{matrix}-i\\i\end{matrix}$	$(x, y);\ (R_x, R_y)$	(xz, yz)

S_6	E	C_3	C_3^{2}	i	S_6^{5}	S_6		$\varepsilon = \exp(2\pi i/3)$
A_g	1	1	1	1	1	1	R_z	$x^2+y^2,\ z^2$
E_g	$\begin{Bmatrix}1\\1\end{Bmatrix}$	$\begin{matrix}\varepsilon\\\varepsilon^*\end{matrix}$	$\begin{matrix}\varepsilon^*\\\varepsilon\end{matrix}$	$\begin{matrix}1\\1\end{matrix}$	$\begin{matrix}\varepsilon\\\varepsilon^*\end{matrix}$	$\begin{matrix}\varepsilon^*\\\varepsilon\end{matrix}$	(R_x, R_y)	$(x^2-y^2,\ xy);\ (xz, yz)$
A_u	1	1	1	-1	-1	-1	z	
E_u	$\begin{Bmatrix}1\\1\end{Bmatrix}$	$\begin{matrix}\varepsilon\\\varepsilon^*\end{matrix}$	$\begin{matrix}\varepsilon^*\\\varepsilon\end{matrix}$	$\begin{matrix}-1\\-1\end{matrix}$	$\begin{matrix}-\varepsilon\\-\varepsilon^*\end{matrix}$	$\begin{matrix}-\varepsilon^*\\-\varepsilon\end{matrix}$	(x, y)	

S_8	E	S_8	C_4	S_8^{3}	C_2	S_8^{5}	C_4^{3}	S_8^{7}		$\varepsilon = \exp(2\pi i/8)$
A	1	1	1	1	1	1	1	1	R_z	$x^2+y^2,\ z^2$
B	1	-1	1	-1	1	-1	1	-1	z	
E_1	$\begin{Bmatrix}1\\1\end{Bmatrix}$	$\begin{matrix}\varepsilon\\\varepsilon^*\end{matrix}$	$\begin{matrix}i\\-i\end{matrix}$	$\begin{matrix}-\varepsilon^*\\-\varepsilon\end{matrix}$	$\begin{matrix}-1\\-1\end{matrix}$	$\begin{matrix}-\varepsilon\\-\varepsilon^*\end{matrix}$	$\begin{matrix}-i\\i\end{matrix}$	$\begin{matrix}\varepsilon^*\\\varepsilon\end{matrix}$	$(x, y);\ (R_x, R_y)$	
E_2	$\begin{Bmatrix}1\\1\end{Bmatrix}$	$\begin{matrix}i\\-i\end{matrix}$	$\begin{matrix}-1\\-1\end{matrix}$	$\begin{matrix}-i\\i\end{matrix}$	$\begin{matrix}1\\1\end{matrix}$	$\begin{matrix}i\\-i\end{matrix}$	$\begin{matrix}-1\\-1\end{matrix}$	$\begin{matrix}-i\\i\end{matrix}$		$(x^2-y^2,\ xy)$
E_3	$\begin{Bmatrix}1\\1\end{Bmatrix}$	$\begin{matrix}-\varepsilon^*\\-\varepsilon\end{matrix}$	$\begin{matrix}-i\\i\end{matrix}$	$\begin{matrix}\varepsilon\\\varepsilon^*\end{matrix}$	$\begin{matrix}-1\\-1\end{matrix}$	$\begin{matrix}\varepsilon^*\\\varepsilon\end{matrix}$	$\begin{matrix}i\\-i\end{matrix}$	$\begin{matrix}-\varepsilon\\-\varepsilon^*\end{matrix}$		(xz, yz)

9. THE CUBIC GROUPS

T	E	$4C_3$	$4C_3^{2}$	$3C_2$		$\varepsilon = \exp(2\pi i/3)$
A	1	1	1	1		$x^2+y^2+z^2$
E	$\begin{Bmatrix}1\\1\end{Bmatrix}$	$\begin{matrix}\varepsilon\\\varepsilon^*\end{matrix}$	$\begin{matrix}\varepsilon^*\\\varepsilon\end{matrix}$	$\begin{matrix}1\\1\end{matrix}$		$(2z^2-x^2-y^2,\ x^2-y^2)$
T	3	0	0	-1	$(R_x, R_y, R_z);\ (x, y, z)$	(xy, xz, yz)

T_h	E	$4C_3$	$4C_3^2$	$3C_2$	i	$4S_6$	$4S_6^5$	$3\sigma_h$		$\varepsilon = \exp(2\pi i/3)$
A_g	1	1	1	1	1	1	1	1		$x^2+y^2+z^2$
A_u	1	1	1	1	-1	-1	-1	-1		
E_g	$\begin{cases}1\\1\end{cases}$	$\begin{matrix}\varepsilon\\\varepsilon^*\end{matrix}$	$\begin{matrix}\varepsilon^*\\\varepsilon\end{matrix}$	$\begin{matrix}1\\1\end{matrix}$	$\begin{matrix}1\\1\end{matrix}$	$\begin{matrix}\varepsilon\\\varepsilon^*\end{matrix}$	$\begin{matrix}\varepsilon^*\\\varepsilon\end{matrix}$	$\begin{matrix}1\\1\end{matrix}$		$(2z^2-x^2-y^2,$ $x^2-y^2)$
E_u	$\begin{cases}1\\1\end{cases}$	$\begin{matrix}\varepsilon\\\varepsilon^*\end{matrix}$	$\begin{matrix}\varepsilon^*\\\varepsilon\end{matrix}$	$\begin{matrix}1\\1\end{matrix}$	$\begin{matrix}-1\\-1\end{matrix}$	$\begin{matrix}-\varepsilon\\-\varepsilon^*\end{matrix}$	$\begin{matrix}-\varepsilon^*\\-\varepsilon\end{matrix}$	$\begin{matrix}-1\\-1\end{matrix}$		
T_g	3	0	0	-1	1	0	0	-1	(R_x, R_y, R_z)	(xz, yz, xy)
T_u	3	0	0	-1	-1	0	0	1	(x, y, z)	

T_d	E	$8C_3$	$3C_2$	$6S_4$	$6\sigma_d$		
A_1	1	1	1	1	1		$x^2+y^2+z^2$
A_2	1	1	1	-1	-1		
E	2	-1	2	0	0		$(2z^2-x^2-y^2,$ $x^2-y^2)$
T_1	3	0	-1	1	-1	(R_x, R_y, R_z)	
T_2	3	0	-1	-1	1	(x, y, z)	(xy, xz, yz)

O	E	$6C_4$	$3C_2(=C_4^2)$	$8C_3$	$6C_2$		
A_1	1	1	1	1	1		$x^2+y^2+z^2$
A_2	1	-1	1	1	-1		
E	2	0	2	-1	0		$(2z^2-x^2-y^2,$ $x^2-y^2)$
T_1	3	1	-1	0	-1	$(R_x, R_y, R_z); (x, y, z)$	
T_2	3	-1	-1	0	1		(xy, xz, yz)

O_h	E	$8C_3$	$6C_2$	$6C_4$	$3C_2(=C_4^2)$	i	$6S_4$	$8S_6$	$3\sigma_h$	$6\sigma_d$		
A_{1g}	1	1	1	1	1	1	1	1	1	1		$x^2+y^2+z^2$
A_{2g}	1	1	-1	-1	1	1	-1	1	1	-1		
E_g	2	-1	0	0	2	2	0	-1	2	0		$(2z^2-x^2-y^2,$ $x^2-y^2)$
T_{1g}	3	0	-1	1	-1	3	1	0	-1	-1	(R_x, R_y, R_z)	(xz, yz, xy)
T_{2g}	3	0	1	-1	-1	3	-1	0	-1	1		
A_{1u}	1	1	1	1	1	-1	-1	-1	-1	-1		
A_{2u}	1	1	-1	-1	1	-1	1	-1	-1	1		
E_u	2	-1	0	0	2	-2	0	1	-2	0		
T_{1u}	3	0	-1	1	-1	-3	-1	0	1	1	(x, y, z)	
T_{2u}	3	0	1	-1	-1	-3	1	0	1	-1		

D_{6h}	E	$2C_6$	$2C_3$	C_2	$3C_2'$	$3C_2''$	i	$2S_3$	$2S_6$	σ_h	$3\sigma_d$	$3\sigma_v$		
A_{1g}	1	1	1	1	1	1	1	1	1	1	1	1		$x^2 + y^2, z^2$
A_{2g}	1	1	1	1	-1	-1	1	1	1	1	-1	-1	R_z	
B_{1g}	1	-1	1	-1	1	-1	1	-1	1	-1	1	-1		
B_{2g}	1	-1	1	-1	-1	1	1	-1	1	-1	-1	1		
E_{1g}	2	1	-1	-2	0	0	2	1	-1	-2	0	0	(R_x, R_y)	(xz, yz)
E_{2g}	2	-1	-1	2	0	0	2	-1	-1	2	0	0		$(x^2 - y^2, xy)$
A_{1u}	1	1	1	1	1	1	-1	-1	-1	-1	-1	-1		
A_{2u}	1	1	1	1	-1	-1	-1	-1	-1	-1	1	1	z	
B_{1u}	1	-1	1	-1	1	-1	-1	1	-1	1	-1	1		
B_{2u}	1	-1	1	-1	-1	1	-1	1	-1	1	1	-1		
E_{1u}	2	1	-1	-2	0	0	-2	-1	1	2	0	0	(x, y)	
E_{2u}	2	-1	-1	2	0	0	-2	1	1	-2	0	0		

6. THE D_{nh} GROUPS (continued)

D_{8h}	E	$2C_8$	$2C_8^3$	$2C_4$	C_2	$4C_2'$	$4C_2''$	i	$2S_8$	$2S_8^3$	$2S_4$	σ_h	$4\sigma_d$	$4\sigma_v$		
A_{1g}	1	1	1	1	1	1	1	1	1	1	1	1	1	1		$x^2 + y^2, z^2$
A_{2g}	1	1	1	1	1	-1	-1	1	1	1	1	1	-1	-1	R_z	
B_{1g}	1	-1	-1	1	1	1	-1	1	-1	-1	1	1	1	-1		
B_{2g}	1	-1	-1	1	1	-1	1	1	-1	-1	1	1	-1	1		
E_{1g}	2	$\sqrt{2}$	$-\sqrt{2}$	0	-2	0	0	2	$\sqrt{2}$	$-\sqrt{2}$	0	-2	0	0	(R_x, R_y)	(xz, yz)
E_{2g}	2	0	0	-2	2	0	0	2	0	0	-2	2	0	0		$(x^2 - y^2, xy)$
E_{3g}	2	$-\sqrt{2}$	$\sqrt{2}$	0	-2	0	0	2	$-\sqrt{2}$	$\sqrt{2}$	0	-2	0	0		
A_{1u}	1	1	1	1	1	1	1	-1	-1	-1	-1	-1	-1	-1		
A_{2u}	1	1	1	1	-1	-1	-1	-1	-1	-1	-1	-1	1	1	z	
B_{1u}	1	-1	-1	1	1	1	-1	-1	1	1	-1	-1	-1	1		
B_{2u}	1	-1	-1	1	1	-1	1	-1	1	1	-1	-1	1	-1		
E_{1u}	2	$\sqrt{2}$	$-\sqrt{2}$	0	-2	0	0	-2	$-\sqrt{2}$	$\sqrt{2}$	0	2	0	0	(x, y)	
E_{2u}	2	0	0	-2	2	0	0	-2	0	0	2	-2	0	0		
E_{3u}	2	$-\sqrt{2}$	$\sqrt{2}$	0	-2	0	0	-2	$\sqrt{2}$	$-\sqrt{2}$	0	2	0	0		

7. THE D_{nd} GROUPS

D_{2d}	E	$2S_4$	C_2	$2C_2'$	$2\sigma_d$		
A_1	1	1	1	1	1		$x^2 + y^2,\ z^2$
A_2	1	1	1	-1	-1	R_z	
B_1	1	-1	1	1	-1		$x^2 - y^2$
B_2	1	-1	1	-1	1	z	xy
E	1	0	-2	0	0	$(x, y),$ (R_x, R_y)	(xz, yz)

D_{3d}	E	$2C_3$	$3C_2$	i	$2S_6$	$3\sigma_d$		
A_{1g}	1	1	1	1	1	1		$x^2 + y^2,\ z^2$
A_{2g}	1	1	-1	1	1	-1	R_z	
E_g	2	-1	0	2	-1	0	(R_x, R_y)	$(x^2 - y^2, xy)$ (xz, yz)
A_{1u}	1	1	1	-1	-1	-1		
A_{2u}	1	1	-1	-1	-1	1	z	
E_u	2	-1	0	-2	1	0	(x, y)	

D_{4d}	E	$2S_8$	$2C_4$	$2S_8^3$	C_2	$4C_2'$	$4\sigma_d$		
A_1	1	1	1	1	1	1	1		$x^2 + y^2,\ z^2$
A_2	1	1	1	1	1	-1	-1	R_z	
B_1	1	-1	1	-1	1	1	-1		
B_2	1	-1	1	-1	1	-1	1	z	
E_1	2	$\sqrt{2}$	0	$-\sqrt{2}$	-2	0	0	(x, y)	
E_2	2	0	-2	0	2	0	0		$(x^2 - y^2, xy)$
E_3	2	$-\sqrt{2}$	0	$\sqrt{2}$	-2	0	0	(R_x, R_y)	(xz, yz)

D_{5d}	E	$2C_5$	$2C_5^2$	$5C_2$	i	$2S_{10}^3$	$2S_{10}$	$5\sigma_d$		
A_{1g}	1	1	1	1	1	1	1	1		$x^2 + y^2,\ z^2$
A_{2g}	1	1	1	-1	1	1	1	-1	R_z	
E_{1g}	2	$2\cos 72°$	$2\cos 144°$	0	2	$2\cos 72°$	$2\cos 144°$	0	(R_x, R_y)	(xz, yz)
E_{2g}	2	$2\cos 144°$	$2\cos 72°$	0	2	$2\cos 144°$	$2\cos 72°$	0		$(x^2 - y^2, xy)$
A_{1u}	1	1	1	1	-1	-1	-1	-1		
A_{2u}	1	1	1	-1	-1	-1	-1	1	z	
E_{1u}	2	$2\cos 72°$	$2\cos 144°$	0	-2	$-2\cos 72°$	$-2\cos 144°$	0	(x, y)	
E_{2u}	2	$2\cos 144°$	$2\cos 72°$	0	-2	$-2\cos 144°$	$-2\cos 72°$	0		

7. THE D_{nd} GROUPS (*continued*)

D_{6d}	E	$2S_{12}$	$2C_6$	$2S_4$	$2C_3$	$2S_{12}{}^5$	C_2	$6C_2'$	$6\sigma_d$		
A_1	1	1	1	1	1	1	1	1	1		$x^2 + y^2,\ z^2$
A_2	1	1	1	1	1	1	1	-1	-1	R_z	
B_1	1	-1	1	-1	1	-1	1	1	-1		
B_2	1	-1	1	-1	1	-1	1	-1	1	z	
E_1	2	$\sqrt{3}$	1	0	-1	$-\sqrt{3}$	-2	0	0	(x, y)	
E_2	2	1	-1	-2	-1	1	2	0	0		$(x^2 - y^2,\ xy)$
E_3	2	0	-2	0	2	0	-2	0	0		
E_4	2	-1	-1	2	-1	-1	2	0	0		
E_5	2	$-\sqrt{3}$	1	0	-1	$\sqrt{3}$	-2	0	0	(R_x, R_y)	(xz, yz)

8. THE S_n GROUPS

S_4	E	S_4	C_2	$S_4{}^3$		
A	1	1	1	1	R_z	$x^2 + y^2,\ z^2$
B	1	-1	1	-1	z	$x^2 - y^2,\ xy$
E	$\begin{Bmatrix} 1 \\ 1 \end{Bmatrix}$	$\begin{matrix} i \\ -i \end{matrix}$	$\begin{matrix} -1 \\ -1 \end{matrix}$	$\begin{matrix} -i \\ i \end{matrix}$	$(x, y);\ (R_x, R_y)$	(xz, yz)

S_6	E	C_3	$C_3{}^2$	i	$S_6{}^5$	S_6		$\varepsilon = \exp(2\pi i/3)$
A_g	1	1	1	1	1	1	R_z	$x^2 + y^2,\ z^2$
E_g	$\begin{Bmatrix} 1 \\ 1 \end{Bmatrix}$	$\begin{matrix} \varepsilon \\ \varepsilon^* \end{matrix}$	$\begin{matrix} \varepsilon^* \\ \varepsilon \end{matrix}$	$\begin{matrix} 1 \\ 1 \end{matrix}$	$\begin{matrix} \varepsilon \\ \varepsilon^* \end{matrix}$	$\begin{matrix} \varepsilon^* \\ \varepsilon \end{matrix}$	(R_x, R_y)	$(x^2 - y^2,\ xy);$ (xz, yz)
A_u	1	1	1	-1	-1	-1	z	
E_u	$\begin{Bmatrix} 1 \\ 1 \end{Bmatrix}$	$\begin{matrix} \varepsilon \\ \varepsilon^* \end{matrix}$	$\begin{matrix} \varepsilon^* \\ \varepsilon \end{matrix}$	$\begin{matrix} -1 \\ -1 \end{matrix}$	$\begin{matrix} -\varepsilon \\ -\varepsilon^* \end{matrix}$	$\begin{matrix} -\varepsilon^* \\ -\varepsilon \end{matrix}$	(x, y)	

S_8	E	S_8	C_4	$S_8{}^3$	C_2	$S_8{}^5$	$C_4{}^3$	$S_8{}^7$		$\varepsilon = \exp(2\pi i/8)$
A	1	1	1	1	1	1	1	1	R_z	$x^2 + y^2,\ z^2$
B	1	-1	1	-1	1	-1	1	-1	z	
E_1	$\begin{Bmatrix} 1 \\ 1 \end{Bmatrix}$	$\begin{matrix} \varepsilon \\ \varepsilon^* \end{matrix}$	$\begin{matrix} i \\ -i \end{matrix}$	$\begin{matrix} -\varepsilon^* \\ -\varepsilon \end{matrix}$	$\begin{matrix} -1 \\ -1 \end{matrix}$	$\begin{matrix} -\varepsilon \\ -\varepsilon^* \end{matrix}$	$\begin{matrix} -i \\ i \end{matrix}$	$\begin{matrix} \varepsilon^* \\ \varepsilon \end{matrix}$	$(x, y);$ (R_x, R_y)	
E_2	$\begin{Bmatrix} 1 \\ 1 \end{Bmatrix}$	$\begin{matrix} i \\ -i \end{matrix}$	$\begin{matrix} -1 \\ -1 \end{matrix}$	$\begin{matrix} -i \\ i \end{matrix}$	$\begin{matrix} 1 \\ 1 \end{matrix}$	$\begin{matrix} i \\ -i \end{matrix}$	$\begin{matrix} -1 \\ -1 \end{matrix}$	$\begin{matrix} -i \\ i \end{matrix}$		$(x^2 - y^2,\ xy)$
E_3	$\begin{Bmatrix} 1 \\ 1 \end{Bmatrix}$	$\begin{matrix} -\varepsilon^* \\ -\varepsilon \end{matrix}$	$\begin{matrix} -i \\ i \end{matrix}$	$\begin{matrix} \varepsilon \\ \varepsilon^* \end{matrix}$	$\begin{matrix} -1 \\ -1 \end{matrix}$	$\begin{matrix} \varepsilon^* \\ \varepsilon \end{matrix}$	$\begin{matrix} i \\ -i \end{matrix}$	$\begin{matrix} -\varepsilon \\ -\varepsilon^* \end{matrix}$		(xz, yz)

9. THE CUBIC GROUPS

T	E	$4C_3$	$4C_3{}^2$	$3C_2$		$\varepsilon = \exp(2\pi i/3)$
A	1	1	1	1		$x^2 + y^2 + z^2$
E	$\begin{Bmatrix} 1 \\ 1 \end{Bmatrix}$	$\begin{matrix} \varepsilon \\ \varepsilon^* \end{matrix}$	$\begin{matrix} \varepsilon^* \\ \varepsilon \end{matrix}$	$\begin{matrix} 1 \\ 1 \end{matrix}$		$(2z^2 - x^2 - y^2,$ $x^2 - y^2)$
T	3	0	0	-1	$(R_x, R_y, R_z);\ (x, y, z)$	(xy, xz, yz)

T_h	E	$4C_3$	$4C_3^2$	$3C_2$	i	$4S_6$	$4S_6^5$	$3\sigma_h$		$\varepsilon = \exp(2\pi i/3)$
A_g	1	1	1	1	1	1	1	1		$x^2 + y^2 + z^2$
A_u	1	1	1	1	-1	-1	-1	-1		
E_g	$\begin{cases}1 \\ 1\end{cases}$	$\begin{matrix}\varepsilon \\ \varepsilon^*\end{matrix}$	$\begin{matrix}\varepsilon^* \\ \varepsilon\end{matrix}$	$\begin{matrix}1 \\ 1\end{matrix}$	$\begin{matrix}1 \\ 1\end{matrix}$	$\begin{matrix}\varepsilon \\ \varepsilon^*\end{matrix}$	$\begin{matrix}\varepsilon^* \\ \varepsilon\end{matrix}$	$\begin{matrix}1 \\ 1\end{matrix}$		$(2z^2 - x^2 - y^2,$ $x^2 - y^2)$
E_u	$\begin{cases}1 \\ 1\end{cases}$	$\begin{matrix}\varepsilon \\ \varepsilon^*\end{matrix}$	$\begin{matrix}\varepsilon^* \\ \varepsilon\end{matrix}$	$\begin{matrix}1 \\ 1\end{matrix}$	$\begin{matrix}-1 \\ -1\end{matrix}$	$\begin{matrix}-\varepsilon \\ -\varepsilon^*\end{matrix}$	$\begin{matrix}-\varepsilon^* \\ -\varepsilon\end{matrix}$	$\begin{matrix}-1 \\ -1\end{matrix}$		
T_g	3	0	0	-1	1	0	0	-1	(R_x, R_y, R_z)	(xz, yz, xy)
T_u	3	0	0	-1	-1	0	0	1	(x, y, z)	

T_d	E	$8C_3$	$3C_2$	$6S_4$	$6\sigma_d$		
A_1	1	1	1	1	1		$x^2 + y^2 + z^2$
A_2	1	1	1	-1	-1		
E	2	-1	2	0	0		$(2z^2 - x^2 - y^2,$ $x^2 - y^2)$
T_1	3	0	-1	1	-1	(R_x, R_y, R_z)	
T_2	3	0	-1	-1	1	(x, y, z)	(xy, xz, yz)

O	E	$6C_4$	$3C_2(=C_4^2)$	$8C_3$	$6C_2$		
A_1	1	1	1	1	1		$x^2 + y^2 + z^2$
A_2	1	-1	1	1	-1		
E	2	0	2	-1	0		$(2z^2 - x^2 - y^2,$ $x^2 - y^2)$
T_1	3	1	-1	0	-1	$(R_x, R_y, R_z); (x, y, z)$	
T_2	3	-1	-1	0	1		(xy, xz, yz)

O_h	E	$8C_3$	$6C_2$	$6C_4$	$3C_2(=C_4^2)$	i	$6S_4$	$8S_6$	$3\sigma_h$	$6\sigma_d$		
A_{1g}	1	1	1	1	1	1	1	1	1	1		$x^2 + y^2 + z^2$
A_{2g}	1	1	-1	-1	1	1	-1	1	1	-1		
E_g	2	-1	0	0	2	2	0	-1	2	0		$(2z^2 - x^2 - y^2,$ $x^2 - y^2)$
T_{1g}	3	0	-1	1	-1	3	1	0	-1	-1	(R_x, R_y, R_z)	
T_{2g}	3	0	1	-1	-1	3	-1	0	-1	1		(xz, yz, xy)
A_{1u}	1	1	1	1	1	-1	-1	-1	-1	-1		
A_{2u}	1	1	-1	-1	1	-1	1	-1	-1	1		
E_u	2	-1	0	0	2	-2	0	1	-2	0		
T_{1u}	3	0	-1	1	-1	-3	-1	0	1	1	(x, y, z)	
T_{2u}	3	0	1	-1	-1	-3	1	0	1	-1		

10. THE GROUPS $C_{\infty v}$ AND $D_{\infty h}$ FOR LINEAR MOLECULES

$C_{\infty v}$	E	$2C_\infty^\Phi$	$\ldots$	$\infty\sigma_v$		
$A_1 \equiv \Sigma^+$	1	1	$\ldots$	1	z	$x^2 + y^2, z^2$
$A_2 \equiv \Sigma^-$	1	1	$\ldots$	-1	R_z	
$E_1 \equiv \Pi$	2	$2\cos\Phi$	$\ldots$	0	$(x, y); (R_x, R_y)$	(xz, yz)
$E_2 \equiv \Delta$	2	$2\cos 2\Phi$	$\ldots$	0		$(x^2 - y^2, xy)$
$E_3 \equiv \Phi$	2	$2\cos 3\Phi$	$\ldots$	0		
$\ldots$	$\ldots$	$\ldots$	$\ldots$	$\ldots$		

$D_{\infty h}$	E	$2C_\infty^\Phi$	$\ldots$	$\infty\sigma_v$	i	$2S_\infty^\Phi$	$\ldots$	∞C_2		
Σ_g^+	1	1	$\ldots$	1	1	1	$\ldots$	1		$x^2 + y^2, z^2$
Σ_g^-	1	1	$\ldots$	-1	1	1	$\ldots$	-1	R_z	
Π_g	2	$2\cos\Phi$	$\ldots$	0	2	$-2\cos\Phi$	$\ldots$	0	(R_x, R_y)	(xz, yz)
Δ_g	2	$2\cos 2\Phi$	$\ldots$	0	2	$2\cos 2\Phi$	$\ldots$	0		$(x^2 - y^2, xy)$
$\ldots$	$\ldots$	$\ldots$	$\ldots$	$\ldots$	$\ldots$	$\ldots$	$\ldots$	$\ldots$		
Σ_u^+	1	1	$\ldots$	1	-1	-1	$\ldots$	-1	z	
Σ_u^-	1	1	$\ldots$	-1	-1	-1	$\ldots$	1		
Π_u	2	$2\cos\Phi$	$\ldots$	0	-2	$2\cos\Phi$	$\ldots$	0	(x, y)	
Δ_u	2	$2\cos 2\Phi$	$\ldots$	0	-2	$-2\cos 2\Phi$	$\ldots$	0		
$\ldots$	$\ldots$	$\ldots$	$\ldots$	$\ldots$	$\ldots$	$\ldots$	$\ldots$	$\ldots$		

11. THE ICOSAHEDRAL GROUPS[a]

I_h	E	$12C_5$	$12C_5^2$	$20C_3$	$15C_2$	i	$12S_{10}$	$12S_{10}^3$	$20S_6$	15σ		
A_g	1	1	1	1	1	1	1	1	1	1		$x^2 + y^2 + z^2$
T_{1g}	3	$\frac{1}{2}(1 + \sqrt{5})$	$\frac{1}{2}(1 - \sqrt{5})$	0	-1	3	$\frac{1}{2}(1 - \sqrt{5})$	$\frac{1}{2}(1 + \sqrt{5})$	0	-1	(R_x, R_y, R_z)	
T_{2g}	3	$\frac{1}{2}(1 - \sqrt{5})$	$\frac{1}{2}(1 + \sqrt{5})$	0	-1	3	$\frac{1}{2}(1 + \sqrt{5})$	$\frac{1}{2}(1 - \sqrt{5})$	0	-1		
G_g	4	-1	-1	1	0	4	-1	-1	1	0		
H_g	5	0	0	-1	1	5	0	0	-1	1		$(2z^2 - x^2 - y^2,$ $x^2 - y^2,$ $xy, yz, zx)$
A_u	1	1	1	1	1	-1	-1	-1	-1	-1		
T_{1u}	3	$\frac{1}{2}(1 + \sqrt{5})$	$\frac{1}{2}(1 - \sqrt{5})$	0	-1	-3	$-\frac{1}{2}(1 - \sqrt{5})$	$-\frac{1}{2}(1 + \sqrt{5})$	0	1	(x, y, z)	
T_{2u}	3	$\frac{1}{2}(1 - \sqrt{5})$	$\frac{1}{2}(1 + \sqrt{5})$	0	-1	-3	$-\frac{1}{2}(1 + \sqrt{5})$	$-\frac{1}{2}(1 - \sqrt{5})$	0	1		
G_u	4	-1	-1	1	0	-4	1	1	-1	0		
H_u	5	0	0	-1	1	-5	0	0	1	-1		

[a] For the pure rotation group I, the outlined section in the upper left is the character table; the g subscripts should, of course, be dropped and (x, y, z) assigned to the T_1 representation.

SELECTED CONSTANTS*

Symbol	Quantity	Value	
		cgs Units	SI Units
k	Boltzmann constant	1.3807×10^{-16} erg K^{-1}	1.3807×10^{-23} J K^{-1}
h	Planck constant	6.6262×10^{-27} erg s	6.6262×10^{-34} J s
c	speed of light in vacuum	2.9979×10^{10} cm s^{-1}	2.9979×10^{8} m s^{-1}
N	Avogadro constant	6.0220×10^{23} mol^{-1}	6.0220×10^{23} mol^{-1}
e	elementary charge	1.6022×10^{-20} emu or 4.8032×10^{-10} esu	1.6022×10^{-19} C (sA)
amu	atomic mass unit	1.6606×10^{-24} g	1.6606×10^{-27} kg
$\beta(\mu_B)$	Bohr magneton	9.2741×10^{-21} erg G^{-1}	9.2741×10^{-24} J T^{-1}
β_e	Bohr magnetin	4.6685×10^{-5} cm^{-1} G^{-1}	
β_N	nuclear magneton	5.0508×10^{-24} erg G^{-1}	5.0508×10^{-27} J T^{-1}
m_e	electron rest mass	9.1095×10^{-28} g	9.1095×10^{-31} kg
m_p	proton rest mass	1.6726×10^{-24} g	1.6726×10^{-27} kg
μ_e	electron magnetic moment	9.2848×10^{-21} erg G^{-1}	9.2848×10^{-24} J T^{-1}
γ_H	proton magnetogmic ratio		2.6752×10^{3} kg^{-1} sA
π		3.14159 26535 89793 23846 26433 83279 50288	

* T stands for Tesla (kg s^{-2} A^{-1}); C stands for Coulomb; S = second; A = ampere; 1 erg = 1 g cm^2 s^{-2}.

CONVERSION FACTORS

$$1 \text{ cal} = 4.184 \text{ J (Joules)}$$
$$1 \text{ Hz} = 6.6262 \times 10^{-34} \text{ J}$$
$$1 \text{ K} = 1.3807 \times 10^{-27} \text{ J}$$
$$1 \text{ kWh} = 3.6 \times 10^{6} \text{ J}$$

Energy Conversion Factors

	erg/molecule	ev	cm^{-1}	cal/mole
erg/molecule	1	6.242×10^{11}	5.036×10^{15}	1.439×10^{16}
ev	1.602×10^{-12}	1	8,067	23,060
cm^{-1}	1.986×10^{-16}	1.2396×10^{-4}	1	2.858
cal/mole	6.949×10^{-17}	4.338×10^{-5}	0.3499	1

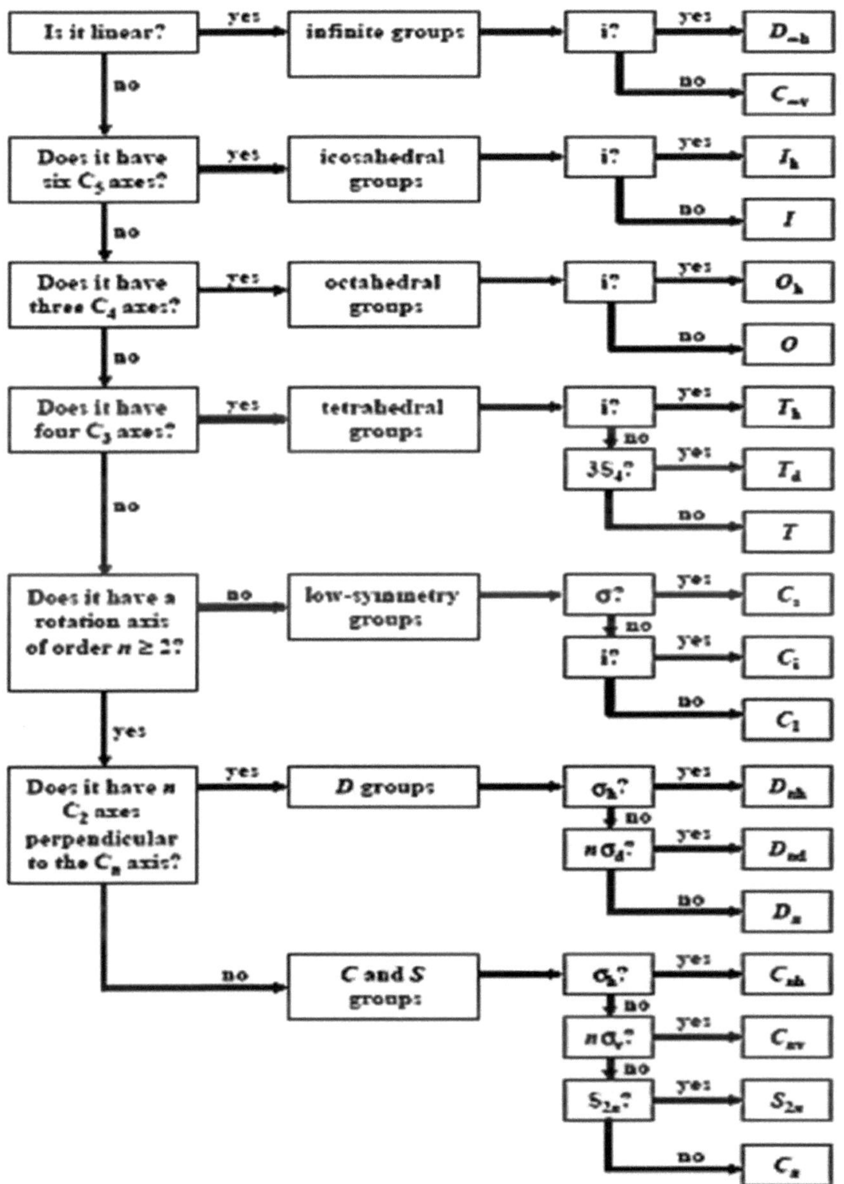

Is it linear?
yes
infinite groups
i?
yes
$D_{\infty h}$
no
$C_{\infty v}$
no
Does it have six C_5 axes?
yes
icosahedral groups
i?
yes
I_h
no
I
no
Does it have three C_4 axes?
yes
octahedral groups
i?
yes
O_h
no
O
no
Does it have four C_3 axes?
yes
tetrahedral groups
i?
yes
T_h
no
$3S_4$?
yes
T_d
no
T
no
Does it have a rotation axis of order $n \geq 2$?
no
low-symmetry groups
σ?
yes
C_s
no
i?
yes
C_i
no
C_1
yes
Does it have n C_2 axes perpendicular to the C_n axis?
yes
D groups
σ_h?
yes
D_{nh}
no
$n\sigma_d$?
yes
D_{nd}
no
D_n
no
C and S groups
σ_h?
yes
C_{nh}
no
$n\sigma_v$?
yes
C_{nv}
no
S_{2n}?
yes
S_{2n}
no
C_n

Table 2.6 Reducible and irreducible representations of C_{3v}

Operations		E	C_3	C_3^2	σ_v	σ_v'	σ_v''
Reducible Represent-tations Γ		$\begin{pmatrix} 1 & 0 & 0 \\ 0 & 1 & 0 \\ 0 & 0 & 1 \end{pmatrix}$	$\begin{pmatrix} -\frac{1}{2} & -\frac{\sqrt{3}}{2} & 0 \\ \frac{\sqrt{3}}{2} & -\frac{1}{2} & 0 \\ 0 & 0 & 1 \end{pmatrix}$	$\begin{pmatrix} -\frac{1}{2} & \frac{\sqrt{3}}{2} & 0 \\ -\frac{\sqrt{3}}{2} & -\frac{1}{2} & 0 \\ 0 & 0 & 1 \end{pmatrix}$	$\begin{pmatrix} \frac{1}{2} & \frac{\sqrt{3}}{2} & 0 \\ \frac{\sqrt{3}}{2} & -\frac{1}{2} & 0 \\ 0 & 0 & 1 \end{pmatrix}$	$\begin{pmatrix} \frac{1}{2} & -\frac{\sqrt{3}}{2} & 0 \\ -\frac{\sqrt{3}}{2} & -\frac{1}{2} & 0 \\ 0 & 0 & 1 \end{pmatrix}$	$\begin{pmatrix} -1 & 0 & 0 \\ 0 & 1 & 0 \\ 0 & 0 & 1 \end{pmatrix}$
Irreducible Represent-ations	Γ_1	(1)	(1)	(1)	(1)	(1)	(1)
	Γ_2	$\begin{pmatrix} 1 & 0 \\ 0 & 1 \end{pmatrix}$	$\begin{pmatrix} -\frac{1}{2} & -\frac{\sqrt{3}}{2} \\ \frac{\sqrt{3}}{2} & -\frac{1}{2} \end{pmatrix}$	$\begin{pmatrix} -\frac{1}{2} & \frac{\sqrt{3}}{2} \\ -\frac{\sqrt{3}}{2} & -\frac{1}{2} \end{pmatrix}$	$\begin{pmatrix} \frac{1}{2} & \frac{\sqrt{3}}{2} \\ \frac{\sqrt{3}}{2} & -\frac{1}{2} \end{pmatrix}$	$\begin{pmatrix} \frac{1}{2} & -\frac{\sqrt{3}}{2} \\ -\frac{\sqrt{3}}{2} & -\frac{1}{2} \end{pmatrix}$	$\begin{pmatrix} -1 & 0 \\ 0 & 1 \end{pmatrix}$
Characters	Γ	3	0	0	1	1	1
	Γ_1	1	1	1	1	1	1
	Γ_2	2	-1	-1	0	0	0

Periodic table

Key to the table:
- Atomic number, Z (top number, e.g. 1)
- Element symbol (e.g. H)
- Relative atomic mass, A (bottom number, e.g. 1.008)

1	2	3	4	5	6	7	8	9	10	11	12	13	14	15	16	17	18
1 H 1.008																	2 He 4.00
3 Li 6.94	4 Be 9.01											5 B 10.81	6 C 12.01	7 N 14.01	8 O 16.00	9 F 19.00	10 Ne 20.18
11 Na 22.99	12 Mg 24.31											13 Al 26.98	14 Si 28.09	15 P 30.97	16 S 32.06	17 Cl 35.45	18 Ar 39.95
19 K 39.10	20 Ca 40.08	21 Sc 44.96	22 Ti 47.90	23 V 50.94	24 Cr 52.01	25 Mn 54.94	26 Fe 55.85	27 Co 58.93	28 Ni 58.69	29 Cu 63.54	30 Zn 65.41	31 Ga 69.72	32 Ge 72.59	33 As 74.92	34 Se 78.96	35 Br 79.91	36 Kr 83.80
37 Rb 85.47	38 Sr 87.62	39 Y 88.91	40 Zr 91.22	41 Nb 92.91	42 Mo 95.94	43 Tc 98.91	44 Ru 101.07	45 Rh 102.91	46 Pd 106.42	47 Ag 107.87	48 Cd 112.40	49 In 114.82	50 Sn 118.71	51 Sb 121.75	52 Te 127.60	53 I 126.90	54 Xe 131.30
55 Cs 132.91	56 Ba 137.34	La–Lu	72 Hf 178.49	73 Ta 180.95	74 W 183.85	75 Re 186.21	76 Os 190.23	77 Ir 192.22	78 Pt 195.08	79 Au 196.97	80 Hg 200.59	81 Tl 204.37	82 Pb 207.19	83 Bi 208.98	84 Po 210	85 At 210	86 Rn 222
87 Fr 223	88 Ra 226.03	Ac–Lr	104 Rf [261]	105 Db [262]	106 Sg [266]	107 Bh [264]	108 Hs [277]	109 Mt [268]	110 Ds [271]	111 Rg [272]	112 Uub [285]						

Lanthanoids

57 La 138.91	58 Ce 140.12	59 Pr 140.91	60 Nd 144.24	61 Pm 146.92	62 Sm 150.35	63 Eu 151.96	64 Gd 157.25	65 Tb 158.92	66 Dy 162.50	67 Ho 164.93	68 Er 167.26	69 Tm 168.93	70 Yb 173.04	71 Lu 174.97

Actinoids

89 Ac 227.03	90 Th 232.04	91 Pa 231.04	92 U 238.03	93 Np 237.05	94 Pu 239.05	95 Am 241.06	96 Cm 244.07	97 Bk 249.08	98 Cf 252.08	99 Es 252.09	100 Fm 257.10	101 Md 258.10	102 No 259	103 Lr 262

Chapter V

Conclusion

A branch of mathematics dedicated to the study of groups is called group theory. The group is one of the most important algebraic structures that plays a central role in abstract algebra and is highly important in various sciences such as crystallography, physics, quantum, and so on.

Suppose a ∈ G and & sigma; (a) = n. If k ∈ Z and ak = e then n | k.

If G is a faraway group.

If G is infinite, then it is uniform with (+, Z).

If the order G is n, then it is uniform with (+, Zn).

The idea of forming group theory came about when mathematicians saw that the structures they study had common properties, and if they could examine all of these properties for a particular structure, they had in fact studied a large part of similar structures and The arrangement saves time. The theory of groups was developed and expanded by major quadrants of classical algebra mathematics, number theory, geometry and analysis. Classical algebra was founded on the polynomial equations by Joseph Louis Lagrange's work in the year 9. Number theory was further studied by Carl Friedrich Gauss in Year 4, and C. F. Klein has done much in the field of geometry and the relationship between geometrical transforms and groups, making him the father of this part of group theory and the founder of the analysis branch Henry Poincaré. Lee Lai and C. F Klein.

References

1. Cotton, F. A., Chemical Applications of Group Theory. Third ed.; John Wiley & Sons: Canada, 1990.

2. Kelly, P., Symmetry. Handout of CHE201. Fall 2008

3. Basic Group Theory, MIT Open Courses 18.904

4. Cotton, F. A., Chemical Applications of Group Theory. Third ed.; John Wiley & Sons: Canada, 1990; Vol. 2, p 8.

5. Kelly, P., Anatomy of a Character Table. Handout Fall 2008.

6. Wiberg, E.; Wiberg, N.; Holleman, A. F., Inorganic Chemistry. Academic Press: 2001; pp. 163-164.

7. Principle of Inorganic Chemistry II, Lecture 6: Molecular Point groups II, MIT Department of Chemistry

yes
I want morebooks!

Buy your books fast and straightforward online - at one of world's fastest growing online book stores! Environmentally sound due to Print-on-Demand technologies.

Buy your books online at
www.morebooks.shop

Kaufen Sie Ihre Bücher schnell und unkompliziert online – auf einer der am schnellsten wachsenden Buchhandelsplattformen weltweit! Dank Print-On-Demand umwelt- und ressourcenschonend produzi ert.

Bücher schneller online kaufen
www.morebooks.shop

Printed by Books on Demand GmbH, Norderstedt / Germany